International Journal on Advances in Life Sciences

2017 vol. 9 nr. 3&4

Printed from e-media with permission by:

Curran Associates, Inc.
57 Morehouse Lane
Red Hook, NY 12571

Printed by Curran Associates, Inc. (2018)

International Academy, Research, and Industry Association (IARIA)
412 Derby Way
Wilmington, DE 19810

Phone: (408) 893-6407
Fax: (408) 527-6351

petre@iaria.org

Additional copies of this publication are available from:

Curran Associates, Inc.
57 Morehouse Lane
Red Hook, NY 12571 USA
Phone: 845-758-0400
Fax: 845-758-2633
Email: curran@proceedings.com
Web: www.proceedings.com

The *International Journal on Advances in Life Sciences* is published by IARIA.
ISSN: 1942-2660
journals site: http://www.iariajournals.org
contact: petre@iaria.org

Sponsored by IARIA
www.iaria.org

International Journal on Advances in Life Sciences
Volume 9, Number 3 & 4, 2017

Eiko Yoneki, University of Cambridge, UK

International Journal on Advances in Life Sciences
Volume 9, Numbers 3 & 4, 2017

CONTENTS

Mario D'Acunto, Institute of Biophysics IBF-CNR, Pisa, Italy
Marco Righi, Institute of Information Science and Technology, ISTI-CNR, Pisa, Italy
Massimo Magrini, Institute of Information Science and Technology, ISTI-CNR, Pisa, Italy
Ovidio Salvetti, Institute of Information Science and Technology, ISTI-CNR, Pisa, Italy

Implementation of OpenEHR in Combination with Clinical Terminologies: Experiences from Norway

Rune Pedersen
Norwegian Centre for eHealth Research
University Hospital of North Norway
Telemedicine- and eHealth
University of Tromsø
Rune.pedersen@unn.no

Conceição Granja, Luis Marco Ruiz
Norwegian Centre for eHealth Research
University Hospital of North Norway
Tromsø, Norway
{conceicao.granja, luis.marco.ruiz}@ehealthresearch.no

Abstract—**Norway is currently involved in several initiatives to adopt clinical information standards and terminologies. This paper aims to identify and discuss challenges and experiences for large-scale national implementation projects when working towards structured Electronic Health Records systems, process and decision support, and secondary use of clinical data. This paper reports from the national strategy for OpenEHR adoption in Northern Norway Regional Health Authority encouraged by the development of a national repository for OpenEHR archetypes and a national initiative to integrate clinical terminologies. The paper contributes to a qualitative longitudinal interpretive study with an effort to increase the possibility to obtain semantic interoperability (towards integrated care) and discusses Systematized Nomenclature of Medicine - Clinical Terms and other relevant clinical terminology and Clinical Information Models such as OpenEHR archetypes. Terminology and archetypes are used to structure the EHR two-folded, and we discuss a general use of information models to increase interoperability extensively. A two-folded use of terminology where terminology is integrated in archetypes, or where terminology is used to structure the Electronic Health Record system while using the hierarchical model of the terminology is discussed. Secondly, we discuss for what purpose OpenEHR is the choice of Clinical Information Model to succeed in Norwegian healthcare. We have identified some challenges and lessons learned.**

Keywords-eHealth medical records; electronic health records; web technology; e-health; interoperability; semantics; integrated care; OpenEHR; terminology; classification systems.

I. INTRODUCTION

Currently, the healthcare system is involved in a transformation process [1][2] in an attempt to overcome some of its challenges [3][4]. Researchers, governments, and international organizations, recognize the need to advance towards a healthcare system capable of integrating the different islands of expertise that conform different components of healthcare [5], accelerate the access to latest evidence [6][7][8], and use the data generated during the care process to elicit new knowledge that allows the whole system to learn from its own experiences [6][9]. Achieving this vision requires an extensive but also sound use of Information and Communication Technologies (ICT) in order to allow healthcare information to seamlessly flow across healthcare levels [6][8]. Achieving this holist healthcare system capable of learning from its own operation needs not only efficient technologies but also the ability of these technologies to exchange information without any ambiguity or loss of meaning, i.e., interoperate at a semantic level [10].

Enabling semantic interoperability (SIOp) across healthcare technological platforms is needed in order to guarantee that different stakeholders will derive the same conclusions from the same data set [10]. If SIOp is not granted across organizational boundaries, the lack of precision in specifying the meaning of the information shared may lead to misinterpretations of healthcare data jeopardizing the quality of care and hampering research outcomes. SIOp is a keystone for holistic healthcare systems that aim to integrate different areas of expertise (and, therefore, the technologies that support them), and reuse their data to generate new knowledge. During the last decades many initiatives have advanced in the development of different standards to enable SIOp [11][12][13]. However, despite the heavy investment performed, standardisation in health care has proven to be a cumbersome and difficult process [14][15].

In Norway, several projects are currently working towards realizing that vision [7][13][16]. For more than a decade, National initiatives towards shared and integrated care have been a focus area for the Norwegian health authorities [17][18]. More recently, several initiatives to enable the secondary use of clinical information were also considered strategic by the Norwegian Health authorities [7][16]. These initiatives especially emphasize the need to apply clinical information standards and terminologies for enabling interoperability across heterogeneous health information systems. However, optimal leverage of all the components needed to enable SIOp is not a trivial task, and it is currently a matter of discussion among academia, implementers, and the Norwegian health authorities.

Three main components are necessary to enable SIOp: a) reference models; b) clinical information models (CIMs) (a.k.a. Detailed Clinical Models), and c) biomedical terminologies [10]. In 2012, Norway opted for the adoption of openEHR as standard for specifying EHR information. The objective was twofold: a) providing a set of robust clinical information models (CIMs) (a.k.a. archetypes) to build the national EHR upon; b) enabling SIOp across different health information systems based on these archetypes.

The adoption of openEHR has been an effort involving many stakeholders at different health trusts and healthcare levels. After 9 years, the number of CIMs defined with openEHR is increasing creating a robust set of information models to build the EHR upon [19]. Regarding the adoption of terminologies, the national strategy has focused on ICD-10 for medical diagnosis and reimbursement, together with a more fragmented use of nursing classifications in different EHR systems for clinical purposes. In 2016, Norway joined the IHTSDO acquiring systematized nomenclature of medicine - clinical terms (SNOMED-CT). When it comes to the adoption of SNOMED-CT the first evaluations performed in collaboration with the committees involved in archetypes definition has already unveiled challenges that need to be overcome in order to fully exploit the potential of the terminology.

This paper describes the national work accomplished to support the openEHR modelling of EHR, and the evolution of systems with focus on the development of clinical value. We describe the work performed in: a) CIMs definition as archetypes by multidisciplinary teams of information architects and clinicians; b) experiences regarding the adoption of different terminologies; and c) the evaluation of the adequacy of adopting SNOMED-CT as reference terminology to annotate CIMs.

The remaining of the paper is organized as follows. The background section introduces the use of CIM and terminologies to enable SIOp, the method section describes the data gathering and synthesis performed through interviews, meetings and active participation with the national comities involved in SIOp technologies adoption. The results section shows first the status and accomplishments of the national initiatives in the definition of CIMs (openEHR archetypes). The second section is devoted to report on efforts in terminologies adoption and the current evaluation of the feasibility in adopting SNOMED-CT. The discussion explains the future challenges and raises important areas that have been identified to be critical in the success for adopting openEHR in combination with clinical terminologies for enabling SIOp.

II. BACKGROUND

A. Clinical Information Models

CIMs [20] are models specified in some clinical information standard to express the schema of clinical information entities processed by Health Information Systems (HIS). CIMs are used to appropriately maintain the consistency of clinical information structures inside a HIS, and to enable semantic interoperability across different systems and organizations. This makes CIMs a basic component for the appropriate management of patient data [10]. Moreover, with recent advances in data reuse strategies, CIMs are also playing an important role in defining the clinical information structures needed for architectures oriented to secondary use of data [21][22][23] and the integration of genetic reports in the EHR [24].

In the last decade, the work of different initiatives to model CIMs is leading to the definition of an extensive catalogue of models publically available. These models can be used to drive the development of HIS. Nowadays, there is a considerable diversity in the standards and approaches available to define CIMs. Examples are openEHR, CIMI, HL7 CDA and HL7 FHIR. Although most editorial teams follow similar steps, there exists no unified methodology or guideline for their definition [20]. The scope of modelling initiatives varies significantly from the local to the international level. For example, the international Clinical Knowledge Managers (CKM) and the Clinical Information Model Initiative (CIMI) define CIMs at an international level; the Norwegian CKM defines them at a national level; and the Intermountain Clinical Element Models (CEMs), were defined at intra-organizational level. The work in parallel of different initiatives has led to semantically equivalent models expressed in different information standards, a.k.a. iso-semantic models. In Norway, the openEHR open standard has been the one adopted by 3 of the 4 health regions. OpenEHR relies on a meta-model (i.e., reference model) and a constraint language (i.e., cADL) to define CIMs (referred to as archetypes in openEHR jargon) [25]. The consistent use of archetypes enables interoperability between different openEHR-based EHRs, as well as efficient reuse of data across different contexts [10][26]. Since archetypes represent a consensus over the data structures to represent clinical information, they need to be defined among the different stakeholders that will rely on the archetype to interoperate. Therefore, their definition must be carried out as a collaborative process among multidisciplinary teams of clinicians and information architects [23][27]. This collaborative definition of archetypes is achieved in web platforms where experts review and publish the archetypes that will be used to define the EHR. These platforms are known as CKMs [28][29][30][31].

In Norway, since 2012, several projects have evaluated the adoption of openEHR as standard for enabling SIOp in secondary healthcare [32]. At the moment, 3 of the 4 existing health regions rely on openEHR to enable SIOp in secondary healthcare. The national initiative that deals with archetypes definition has gradually gained a foothold in the Norwegian e-Health scene. The openEHR architecture has been used to build a national CKM. Archetypes are defined collaboratively in the CKM in order to provide vendors a library of common formal models to build their clinical information systems on. The national CKM archives information about how new archetypes are translated, modelled, and shared, and is planned to contain between 1000 to 2000 archetypes. The final aim is to build an open source repository of clinical content, based on the OpenEHR clinical information model. A precondition for success is that clinicians agree on the content of each archetype in the CKM consensus processes. Clinicians from the four Regional Health Authorities are active contributors in the process for developing archetypes. The national editorial group, and the National Administration Office of Archetypes (NRUA, from the Norwegian nasjonalt redaksjonsutvalg for arketyper) have coordinated this process.

The national CKM in Norway is responsible for the definition of reference archetypes that vendors will use to build the EHR information model on. The definition of CIMs typically encompasses two main tasks: a) the specification of the information structure in a clinical information standard such as OpenEHR; and b) the binding of the meaningful sections of the CIM to a terminology to attach unambiguous standard descriptions to them.

Archetype sections can be bound to concepts provided by standard terminologies, thus endowing archetype elements with semantics provided by standard terminologies such as ICD 10, LOINC or SNOMED-CT. This enables semantic interoperability among those HIS that rely on the same set of archetypes and terminology [33]. At present, the binding of the archetypes in the National CKM to SNOMED-CT is under evaluation. The results section reports some of the challenges found to coordinate the use of SNOMED-CT in combination with archetypes.

B. *Biomedical terminologies*

Clinical terminologies offer a common vocabulary for national health authorities, local researchers, and quality registers [34]. This means that in addition to being a storing device for free text data, EHRs are capable of encoding commonly occurring data using fixed lists of multiple choices for certain purposes. Thus, data becomes more comparable and computable than free text would be.

There is a wide spectrum of clinical terminologies. In nursing, the International Classification of Nursing Practice (ICNP) or Nursing Intervention Classification (NIC) and The North American Nursing Diagnosis Association (NANDA) are widely used [27][35]. Other examples are the International Statistical Classification of Diseases and Related Health Problems (ICD), the Foundational Model of Anatomy (FMA) or even general-purpose terminologies such as SNOMED-CT. There are also terminological standards for more specific domains, such as the International Classification of Functioning, Disability and Health (ICF) for rehabilitation. These terminologies have been developed, and used, to ensure consistency of meaning across time and place. On one level, nursing classifications enable day-to-day planning for local users (primary use) where clinical terminology is used to structure information (standardized care-plans) using diagnosis and interventions from NIC and NANDA, for example. In practice, the adoption of standard terminologies will enable a consistent vocabulary to describe information that is sent and received between different systems or health care deliverers. Although SIOp is the main objective, other areas benefit from the consistent use of terminologies and CIMs. The increased focus on process-oriented systems across different health care organisations presumes standardization in the form of shared terminologies and information models to enable SIOp. However, leveraging the use of CIMs and terminologies optimally with CIMs involves challenges that information architects must face at design time [36]. Examples of these challenges include decisions about what to represent as CIMs or terminology, which sections of the archetype to annotate, whether or not to use post-coordination to express clinical concepts, among others.

Archetypes can be tagged with SNOMED-CT codes adding a standard term to each of the sections and nodes of the archetype. This allows specifying the sections and contents of the archetype in a standard terminology so it can be interpreted over organisational or even trans-national borders. In this way, the information becomes interoperable for multiple purposes. Thus, it is essential that standardized terminologies for different domains can be integrated either in the archetype or in the EHR system. The use of a terminology such as SNOMED-CT, which is widely exploited, increases the semantic interoperability at several levels, both for primary and for secondary use. For example, the use of terminologies such as SNOMED-CT facilitates the integration of disparate systems by providing a common definition of clinical terms that can be used to determine when sections of different information models represent the same entity. The integration with co-existing EHR systems is especially important. Medication, laboratory results, and the care plan have to be integrated in the same view to visualize the pathway. The chart systems have a CIM for structured data elements that differs from the OpenEHR/archetype CIM intended to be used in Norway, and mapping between them demands unknown resources.

III. Materials And Methods

The research presented herein has mainly been developed in the North Norwegian Regional Health Authority in coordination with NRUA, and the Norwegian Directorate of Health. Interpretive and ethnographically oriented qualitative methods have been applied, grounded in the participation and contribution to the work accomplished [19]. The fieldwork has focus on the regional/national work accomplished, and, secondarily, on the forthcoming process where numerous archetypes will be tested as structured elements in the new process oriented EHR system. A mind map of the archetype problem/diagnosis is shown in Figure 1. During the last seven years several meetings, courses, and workshops with focus on archetypes and terminology have been covered by observations, document analysis, and interviews. Conversations, discussions, reflections, and debates from these meetings are the foundation of this work. The observations and description of on-going work have been followed by interviews with members of the regional and national initiatives. This includes six interviews on the archetype governance, 10 interviews and 180 hours of observations on the use of clinical terminology, conversations with end users of the CKM while guiding them to become users, and participants in national discussions on the consensus of archetypes. The interviews includes physicians and nurses that are active users of the CKM. The process of educating them to become CKM users has provided valuable knowledge on how to develop the learning and recruitment strategies. A summary of the data collection for the qualitative research process is presented in TABLE I.

TABLE I. An overview of the data collection. Time used, and the number for interviews of the qualitative research process.

Data source	
Interviews with contributors to the work with archetypes, and the development of new EPR.	18 open ended interviews
Participatory observation	180 hours
Participation in meetings, workshops, and informal discussions.	300 hours
Research on SNOMED compared to CIM	200 hours
Document studies: Documents from the CKM, concerning archetypes in general.	

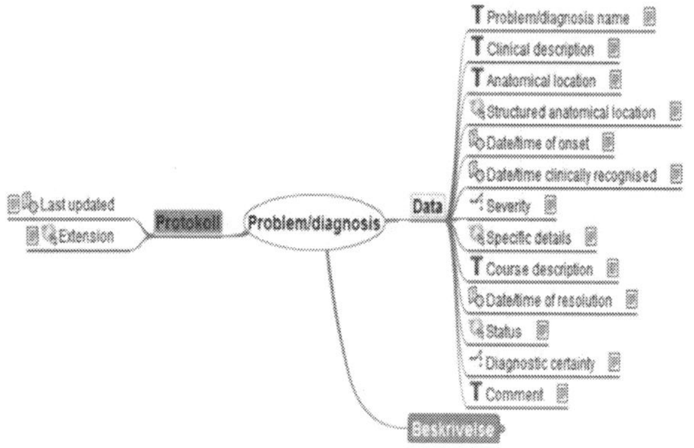

Figure 1. Illustration of the archetype Problem/Diagnosis in the form of a mind map.

IV. Results

A. Definition and governance of archetypes

During the last three years, the use of OpenEHR archetypes has grown with focus on a national anchorage in Norwegian healthcare. The initiative has developed through National ICT, and an EHR vendor that holds more than 80 % of the secondary healthcare EHR systems. From the outset, a national collaborating group is working, in coordination with the aforementioned vendor, to build a national repository of archetypes.

The National ICT, with the goal of producing high quality archetypes, established NRUA in 2013. NRUA has assigned six full/part time associates with an increasing number of collaborators in the Regional Health Authorities. There are between two and three members from each of the four regions. As an example, there is an increasing number of members from the North Norwegian Health Authority, one physician with special interest in health informatics, one nurse with a PhD in Information Systems, one ICT-advisor, two PhD students in part time positions and one project manager from the regional ICT development program where the new process oriented EHR is developed. NRUA also cooperates closely with global connections such as the international openEHR CKM [28], and vendors that cooperate with the Norwegian vendor. The vendors are important contributors with a mutual interdependency. In all, the governance work is important in local, national, and global environments. The overall goal with NRUA is to coordinate the development, and use of archetypes, on a national level, both handling translations of international archetypes as well as handling local initiatives.

The reviews are initiated by the Editorial Group, which also covers the recruitment of the reviewers to the national CKM. The Editorial Group does the further approval if the requirements are met.

The number of review iterations varies depending if the archetype was mature when imported into the Norwegian CKM, or it had to be developed from scratch.

The collaboration and coordination between the national and international CKMs is crucial and, as shown before, helps to reduce the number of review iterations needed to publish an archetype. In some cases, archetypes are imported from the international CKM into the national and published there; while in other cases, the national CKM develops an archetype needed by Norwegian implementers and later it is adopted by the international CKM. Weather archetypes are approved first on international or national CKM depends on the priories of each editorial team. The priority scale is provided in TABLE II. The scale gives an total score for each information element with a total score to coordinate the consensus work. After publication, the results of the review are collated and taken to the other CKM to accelerate the review process. One of the leaders of the international CKM stated, "*the collaboration between the international and the Norwegian CKM is one of a kind based on the national consensus process, and all activities with archetypes in Norway are followed by the international society and vice versa*". Nevertheless, since archetypes publication involves the consensus of many stakeholders with very different backgrounds, there are challenges in the coordination of reviews, and final agreement on the published archetype. She continued by saying, "neither the CKM nor the consensus process is perfect and adjustments will be necessary along the way. Changes can be related to open-source and Web based CKM/process where everything is stored open and is constantly evolving". In relation to these challenges, Christensen and Ellingsen [32] performed an evaluation in the openEHR adoption process between 2015 and 2016 identifying several organizational problems. In their study, they found that the fact of establishing the archetype editorial team (NRUA) at the same time as the implementation of the openEHR-based system that had to use those archetype originated problems. The reason is that the archetypes need to be in a published state when the EHR development starts. Otherwise, the vendor does not count on the real use case models during the implementation. One of the causes for this overlap of activities was the time needed to build a consolidated editorial team with representative reviewers had been underestimated.

TABLE II. Priority scale for archetypes [37]. A score made to differentiate archetypes in process towards consensus. The score 4-15 is a means to prioritize. wiki.arketyper.no (12.01.2016)

Reuse (0-3)			**Diffusion (1-3)**	**Time for testing (1-5)**	**Functional dependencies (1-2)**	**Semantic dependencies (1-2)**	**Total score (4-15)**
Search, presentation, reuse, aggregation	***Process support***	***Decision support***					
1	1	1	3	5	2	1	14

Clinicians are willing to participate in the review of archetypes but often they require that some of their working hours are reallocated for this task [32]. However, these hours are often difficult to be released since they do not depend on NRUA. A clear alignment with local leaders becomes paramount so they understand the benefits of openEHR adoption and they provide some resources in the form of review hours. In the initial stages these organizational factors caused the publication of archetypes to be slow, in fact only one archetype had been published in the first year of work of the Norwegian CKM [32]. Despite these challenges, the following years become more fruitful. During 2015, 2016 and 2017 the review process has become more mature and effective in the publication of archetypes [19]. At present, it counts on a good pool of reviewers and collaborators that have contributed to accelerate the publication of archetypes, see TABLE III. While at the beginning of 2014, NRUA first had focus on the translation of already existing archetypes and observation-archetypes such as blood pressure, body weight, nutritional risk, height, and temperature, at the time of writing, national consensus has been reached for 57 archetypes and more than 100 are in process of approval. Examples are the archetypes:

- openEHR-EHR-OBSERVATION.nutritional_risk_screening.v1;
- openEHR-EHR-OBSERVATION.blood_pressure.v1:
- openEHR-EHR-OBSERVATION.body_weight.v1.

During 2015 and 2017 more complex archetypes were defined such as:

- openEHR-EHR-INSTRUCTION.medication_order.v1;
- openEHR-EHR-EVALUATION.adverse_reaction_risk.v1.

Clinicians have been invited to participate through the national CKM after coordination between the regional groups and the secretariat at NRUA. The CKM is, as showed in Figure 2. , where the clinicians state there opinion in the text box on the right hand side. Archetypes are used as standards for the clinical content of the EHR and it was important for clinicians to have an essential role in defining and designing them. One clinician said: "It is crucial to include clinicians in this work; they have the clinical knowledge and know what is important to focus on, for the archetypes to be useful standards for clinical work." The same clinician commented, "If others than clinicians design the archetypes, it will be troublesome to get clinicians to accept and use them". However, as reported by Christensen and Ellingsen [32], it was not only enough to count on clinical reviewers, those reviewers needed to be trained to review archetypes using the Norwegian CKM.

TABLE III. *DISTRIBUTION OF ARCHETYPES IN EACH OF THE DIFFERENT LIFECYCLES:* PUBLISHED, IN REVIEW OR DRAFT.

Draft	149
Review	26
Published	58

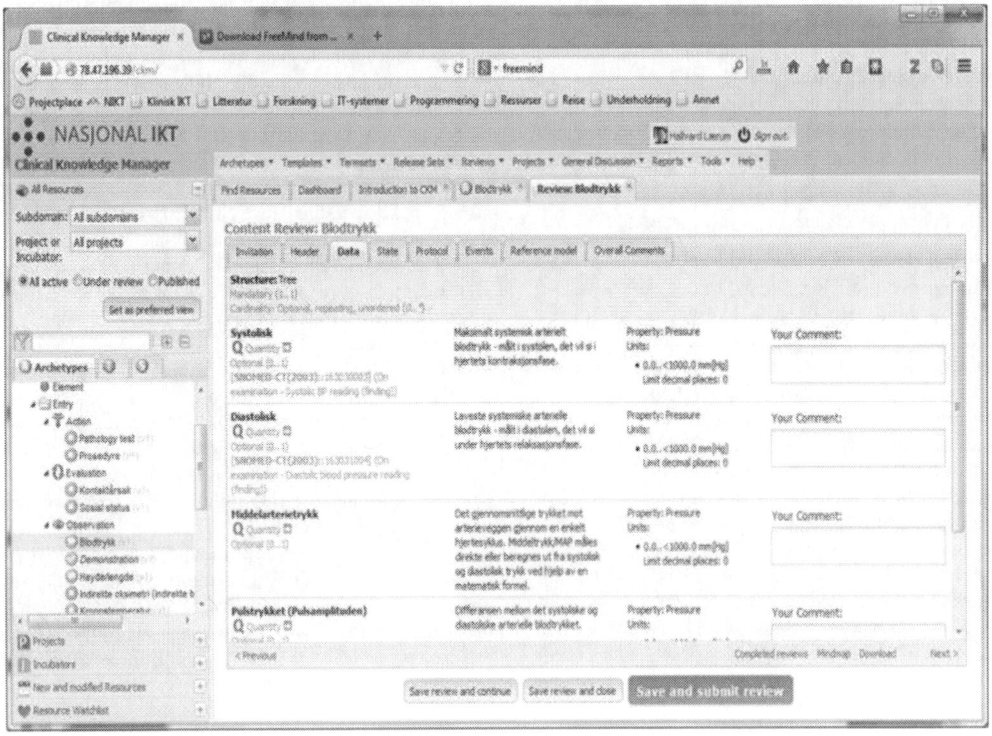

Figure 2. Screenshot of the Clinical Knowlwdge Manager where clinicians cooperate towards the consensus of archetypes.

B. The use of terminology to standardize local practice

In parallel to the adoption of openEHR as information standard, the Norwegian Directorate of Health has put focus on clinical terminology, and has engaged clinicians nationwide to explore the integration of several terminologies in the existing ICT portfolio.

Since 2005, one of the largest hospitals in Norway, Akershus University Hospital, has used an EHR that includes a module for nursing. Along the lines of standardization, the nursing care plan, including nursing classification systems were regarded as a mean for making nursing work more effectively and offering quality assurance. The classification systems are ICT-based standards integrated with the care plan. The diagnoses are represented by the international classification system NANDA, consisting of 206 nursing diagnoses [38]. The interventions are represented by the NIC system, consisting of 486 interventions. Care plans are increasingly made to replace the use of free text in the documentation, foremost to establish a common, formalized language based on the best practices. Free text documentation is whatever information nurses share about the patient in the EHR in addition to, or without, writing formalized care plans. However, the implementation of the EHR led to a systematic use of standardized care plans that gave more efficacy, transparency, and quality of documentation. The care plan has been organized in such a manner that each diagnosis, dimension, and action is firmly attached to the plan with a start and a stop date. When standardizing these plans, the nurse can easily choose several actions from a predefined list for the applicable diagnosis. By doing this, the nurse saves time, while the standardized sentences work as a quality indicator. The purpose of using terminology as a primarily means to standardized EHR systems is challenging, still terminology has been used to structure an unstructured EHR system with success.

C. The national strategy for clinical terminologies

At a national level several terminologies have been adopted and integrated in EHR systems. In the case of ICD-9 and ICD-10, they have been used in Norwegian healthcare systems from their origin with the target of coding medical diagnosis both for clinical and economical benefit. One clinical IT manager stated that the primary target behind the adoption of terminologies could have been achieved if the clinicians had used the diagnosis codes from ICD-10 to categorize the patients in the EHR. However, clinicians found it difficult to be explicit and specific early in the trajectory of the patient, since diagnosis change throughout the patient pathway. Diagnosis change and the IT systems in use need to track and categorize these changes logically to support the activity coding of clinical work. The coding of activities reflects the focus of the clinical pathway, not the diagnosis of the patient. In this sense, the adoption of archetypes becomes valuable: as a quality assurance of the completeness of the clinical terminology, to direct the clinical content of the EHR systems and other integrated systems, and to identify relevant information and give the clinicians access to this information. However, in opposition to local level, patient pathways fixed to clinical ICD diagnosis probably need to be determined on a national level, and based on national directives. This led to discussions on how to use different terminologies in combination with archetypes. A member of the regional archetype group stated: "*The archetype is not annotated but this is a subset of the SNOMED concepts available for severities. As a maximum data set, the archetype should not restrict the "standard" set of terms agreed in terminologies. However, before doing so, I think that the implications in terms of SNOMED licenses should be considered very carefully.*"

The use of standard terminologies is also considered paramount for the standardization of clinical processes. For instance, clinical pathways for cancer diagnosis are today organized from national cancer groups that have resulted in national guidelines that easily could be followed and connected to already existing patient pathway processes, standardized packages to monitor that cancer patients receive the right treatment at the applicable time. Large-scale Infrastructure projects, with increasingly more focus on integrated care, put pressure on the Norwegian Directorate of Health to focus on clinical terminologies and archetypes. Recently, there has been a growing activity in the section of e-health towards increased focus on general-purpose terminologies such as SNOMED-CT, and the International Classification of Functioning, Disability and Health (ICF). This has motivated studies on how terminologies and archetypes fit together. A selected number of experts have, through the last 6 months, recruited evaluators from all over the country. These are clinicians and information architects with special interest in the use of clinical terminology. The work started in November 2015 with the purpose to map SNOMED-CT towards the most commonly used EHR functions. At the same time ICNP will be piloted in the primary healthcare services, this has been organised by the Norwegian Nursing Association that has translated the terminology, and the Norwegian Directorate of Health acknowledges it. SNOMED-CT and ICNP are both discussed in the new national project. Other Scandinavian countries such as Sweden and Denmark have earlier allocated significant resources both to translate and get SNOMED-CT operational for clinical practice.

The national projects have focus on SNOMED-CT and attempt to advise the Ministry of Health in questions such as: should Norway become an organized users of SNOMED? How is the coverage of SNOMED-CT for the content of the clinical pathway? How is the integration of SNOMED-CT solved technically? The last question includes the use of archetypes, but also the possibility to use SNOMED directly to represent EHR content?

A national project coordinated by the Norwegian directorate of Health in 2016 had the purpose to map the clinical patient pathway with SNOMED-CT codes. A standardized breast cancer process was used with the purpose to categorize the coverage of SNOMED-CT for all the relevant clinical variables. The study revealed that despite the total number of codes in the SNOMED-CT hierarchies, more than 30 % of the breast cancer process values had no SNOMED-CT code.

D. Terminology binding of archetypes

A key aspect of archetypes is that they allow the binding of their elements to codes from an external terminology. Since the objective in Norway is to cover most clinical areas in secondary healthcare using archetypes, the adoption of SNOMED-CT as general-purpose terminology for their terminology binding has been explored. The binding of archetypes to SNOMED-CT endows them with the semantics provided by the standard terminology. This allows those systems relying on the same archetypes and SNOMED-CT to interoperate at a semantic level. Concerning the secondary use of clinical information, the adoption of SNOMED-CT facilitates the representation of clinical content with rich semantics enabling expressive queries required for phenotyping in clinical research.

In collaboration with NRUA, the combination of SNOMED-CT with archetypes has been evaluated finding several challenges [39]. At a technical level archetypes can be bound to any terminology and it is up to the archetype designer what elements of it to annotate. However, in our experiences assessing archetypes binding to SNOMED-CT, several challenges were found. Archetypes are generic data schemas to be used at a national level, however their binding to terminologies is very influenced by the different scenarios where they is used. For example, sharing the patient summary across different health platforms would require the annotation of some main sections of the archetype with SNOMED-CT. Guidelines for terminology binding can be followed at that level [36][40][41]. However, if a higher level of expressivity is needed, for example to be able to use the archetype to perform semantic queries applying reasoning over the terminology, more sections of the archetype would need to be linked to SNOMED-CT. It would be required to represent some contextual properties at a terminological level for dealing with contextual aspects such as time intervals, parties involved in an observation etc. For example, if one needs to perform a query for identifying patients diagnosed with some kind of malignant epitherial neoplasms in the last 10 years, any patient diagnosed with a subtype of such disorder should be retrieved, i.e. patients diagnosed in the last decade with adenocarcinoma of nasopharynx, carcinoma of lingual tonsil, carcinoma of the uvula etc. To enable this kind of expressive queries, first, it is required to annotate archetypes and their instances properly with an ontology-based terminology; second, it is necessary to rely on a technical infrastructure that explores the SNOMED-CT concept hierarchy; and, third, timing aspects expressed at an archetype level need to be also be considered in the query processing. SNOMED-CT covers the first two requirements but, as domain ontology, specifying precise contextual aspects such as time is out of its scope [42]. This involves that two different models (the archetype and the terminology) need to be analysed to answer such a query. The terminology can provide subsumptive (i.e. class- subclass) reasoning to identify subtypes of the disorder and the archetype query language (AQL) can help in filtering contextual aspects. However, this means operating at two different levels as proposed in [43] thus making more complex and less dynamic the definition of queries over clinical information. This challenge is not only present when querying the EHR for research purposes. In fact, the presence of iso-semantic models (CIMs defined in different standards) requires the use of semantic web architectures to guarantee that the meaning of health information is preserved across organizational boundaries to provide access decision support systems [44]. How to leverage archetypes and biomedical ontologies such as SNOMED-CT guaranteeing scalability and viability of health information architectures is currently a matter of concern for the Norwegian e-Health Directorate.

V. Discussion

A. Adoption clinical information model

Terminology standards are used on a daily basis in health care work. The combination of health information standards and terminologies is currently the preferred option to enable SIOp for sharing clinical data both for care delivery and for secondary uses (e.g., clinical research) [3]. In Norway, the availability of a structured EHR model based on archetypes is opening the door for clinicians to categorize variables for building meaningful reports, extracting data for quality registers, and performing clinical research. Structured data elements will also make it possible to organize information that supports processes and decision support inside an integrated EHR portfolio and the use of OpenEHR will provide clinicians with a more open, adaptive, and collaborative system. OpenEHR compliant data tagged with clinical terminology codes allows the interoperation of different HIS, thus enabling integrated care. In the near future, the implementation of the archetype-based system will elevate the possibilities to use the standardized information models in clinical settings. The implementation program for the new EHR has focus on clinical decision support, and mapping of CIM between different integrated systems.

B. Primary use of terminology

The integration of clinical terminology for use in EHR systems to support clinical practice has proven difficult to accomplish. With the use of archetypes, and a national governance of clinical variables through a common repository for structured data elements, there are future advantages of both semantic and interoperable character. Earlier research elaborates on how the categorical use of clinical terminology to structure nursing diagnosis and interventions in standardized nursing plans has been a success for increased quality and efficiency. However, the use of clinical terminology to categorized clinical documentation for enabling process- and decision support in the EHR portfolio is limited. The use of standardized nursing plans at a large scale in a Norwegian hospital showed clear advantages for both quality and efficiency [45]. Furthermore, when information is tagged with the purpose to categorize such as with ICD-10 and medical diagnosis, the same information becomes available for secondary purposes. On the next level, any of these tagged nodes of information could be recirculated. Archetype based elements such as blood pressure, pulse, temperature, and laboratory data can also be used for primary purposes. Different national and regional initiatives will in a close future be piloted in different implementation processes such as Helseplattformen that is a pilot for One- citizens one Journal, and the new FIKS program in the North Norwegian Health Authority [46].

C. Secondary use of terminology

At the same time, as the primary information becomes interoperable, both as single archetype/terminology or intervened, the information becomes semantically interoperable for use in secondary settings. As an illustration, all information that is tagged with the nursing classification ICNP, both diagnosis and interventions, becomes sharable for secondary use. All the nursing diagnosis and interventions would be an object for clinical research on a national or global level, which is a relatively unexploited research arena. The care plan is intended to be an interdisciplinary tool for categorizing documentation in the EHR. For this to become a success, it is important that structured information from other applications is used in the care plan. A regional implementation program will soon have focus on the integration of information models between the EHR and the EMR [46].

In the end, an increasing number of archetypes, a so far unknown number, is expected to be accessible in an open repository, and each archetype that is translated or modelled will be compared or reviewed with the purpose of being added to the global repository. For instance, the process of getting consensus on the observation archetype blood pressure started with a translation of the global standard, and ended with a new version that also is planned for the global or international repository. The increasing number of archetypes increases the possibility to use the structured data elements to extract data for other secondary purposes. The national quality registers is an example where physicians use time on extracting data form the HER manually. The same implementation program will focus on possibilities to make these processes automated using archetypes.

D. Adoption clinical information model

The diversity in standards, scope and methodology complicates the decision about adopting one standard or another for the definition of CIMs since it will influence the systems that can be deployed in the health network. Standards available are openEHR, ISO 13606, HL7 CDA, HL7 FHIR, to name a few. In Norway, although 3 of the 4 health regions rely on openEHR for secondary care. Nevertheless, many other HIS rely on different standards or no information standard at all. Therefore, in the near future it is expected to find an ecosystem where implementations based on different information models will coexist. This may add a burden for those implementers that need to adapt from one standard to another. However, it is important to notice that the most valuable resource of a CIM is not the technical specification, but the conceptual model that it contains. CIMs, beyond providing a format to express clinical information, define a way of combining clinical concepts together to build more complex conceptual structures. For example, the archetype OpenEHR-EHR-CLUSTER.symptom_sign.v1 aggregates several granular concepts such as Body site, Episodicity, Impact, etc., to build the more complex entity Symptom/Sign. This aggregation of concepts is more evident when the CIM is annotated with an international terminology. Reaching a consensus about the conceptual model of the CIM is the task that consumes most of the efforts of editorial teams since they need to coordinate professionals from different domains. Nevertheless, if the modelling work is appropriately performed, the conceptual model will be equivalent in most iso-semantic models. Consequently, once CIMs are defined in a particular standard, the conceptual model is clear and can be transformed to other representations/standards. In fact, that is the approach of the opencimi.org initiative, which pursues the definition of CIMs that can be expressed in several formats such as CEMs, HL7 CDA or OWL by defining transformation functions among them. These transformations, although complex, are technical tasks that can be accomplished with much less effort than the definition of stable conceptual models. The EU project SemanticHealthNet has provided insights to define an ontology based on the CIM conceptual model that allows the access to equivalent information hosted in repositories expressed with disparate information standards [47]. In the case of Norway, we expect several standards to be implemented in different regions and HIS. In such scenario, we believe that the common repository of published CIMs expressed as archetypes will be of paramount importance to provide the set of models approved at a national level that can drive the development of interoperable infrastructures across disparate HIS.

VI. Conclusion

Currently several standards and terminologies are available for the specification and annotation of CIMs, respectively. openEHR, HL7 CDA and ISO 13606 are examples of standards to define CIMs, which in most cases are annotated with standard terminologies to enable their interoperability across systems.

With the parallel national initiatives running at this time in different countries, it is starting to become visible how the organization and size of countries influences their standardization efforts. On one hand, large countries with very heterogeneous health networks are aiming for the adoption of standards that allow sharing EHR information documents extracts That is the case of Spain with ISO 13606 [48] or the US with HL7 CDA [12][49].

On the other hand, Norway is heading to the adoption of a nationwide EHR information architecture with openEHR that defines not only some relevant CIMs but also the whole EHR information structure. Three factors have influenced this direction of work. The first is the homogeneity in the market since only one vendor represents 80% of the market share in hospital. The second is the close collaboration between vendors and health authorities; this allows coordinating the definition of the whole information model of new systems. The third, and most determinant, is the body of knowledge already available in the international CKM that has fed the national CIMs definition pipeline with existing archetypes. This has accelerated their validation at a national level avoiding their definition from scratch.

At the moment, the Norwegian eHealth strategy has established a multidisciplinary community of vendors, governmental agencies and health organizations collaborating in order to define a nation-wide EHR information architecture. The knowledge management framework of OpenEHR supports to manage the national CKM. The OpenEHR governance model and the collaboration between the international and Norwegian CKM teams are proving to be effective to manage the definition of CIMs for the national eHealth strategy. On the technical side, the rich reference model provided by OpenEHR acts as a powerful modelling tool for the definition of CIMs. On the organizational side, the collaborative environment provided by the CKM is allowing to ensure the validity of the CIMs generated. As a result, the National eHealth Department is providing the health informatics community a body of standard clinical models, which allows implementers and researchers to define standard interoperable implementations on them.

The semantic interoperability gained from the use of both terminology and OpenEHR archetypes separately is a highly valuable asset. For instance, earlier studies in Norway have showed that clinical terminology has the potential to structure information of unstructured EHR systems. The ongoing national work also suggests that the combined use of archetypes and terminology further increases the semantic interoperability for connecting EHR systems on different layers of healthcare. Using for instance SNOMED "non-hierarchical" to tag the nodes of archetypes is interesting, and could be an integration advantage for vendors. It is a fact that both subjects complement each other's capacity to reach semantic interoperable.

Another "feature" that could increase the semantic interoperability is the growing possibility to use different Clinical Information Models to extract and share information from the National repository of archetypes/ clinical variables and content. The Government and the National e-health administration has decided to use different ICT systems in the primarily and specialist healthcare for several years to come. This requires a possibility to use clinical content from the national repository using another CIM specification standard than OpenEHR to extract and use semantic interoperable information. In this sense, the clinical model defined by an archetype can be represented in another standard by defining transformation rules among OpenEHR and the other standard. This way, the archetype-based repository becomes the reference common information 'ontology' or conceptual model used by different vendors regardless the standard, classes, and model they implemented.

ACKNOWLEDGEMENT

Thanks to the people at North Norwegian Health Authority, the NRUA, and the National ICT for contributing with vital information to the project.

REFERENCES

[1] R. Pedersen, C. Granja, and L. M. Ruis, "The Value of Clinical Information Models and Terminology for Sharing Clinical," The Eighth International Conference on eHealth, Telemedicine, and Social Medicine (eTELEMED 2016), Venice, Italy, IARIA, 2016, pp. 153-159.

[2] C. P. Friedman, A. K. Wong, and D. Blumenthal, "Achieving a Nationwide Learning Health System," Science translational medicine, vol. 2, pp. 57cm29-57cm29, 2010.

[3] T. Greenhalgh, J. Howick, and N. Maskrey, "Evidence Based Medicine: A Movement in Crisis?," Bmj, vol. 348, p. g3725, 2014.

[4] B. Djulbegovic, G. H. Guyatt, and R. E. Ashcroft, "Epistemologic Inquiries in Evidence-Based Medicine," Cancer control: journal of the Moffitt Cancer Center, vol. 16, pp. 158-168, 2009.

[5] M. Rigby, P. Hill, S. Koch, and D. Keeling, "Social Care Informatics as an Essential Part of Holistic Health Care: A Call for Action," International Journal of Medical Informatics, vol. 80, pp. 544-554, 2011.

[6] C. Friedman and M. Rigby, "Conceptualising and Creating a Global Learning Health System," International journal of medical informatics, vol. 82, pp. e63-e71, 2013.

[7] A. Budrionis, L. Marco-Ruiz, K. Y. Yigzaw, and J. G. Bellika, "Building a Learning Healthcare System in North Norway," Proceedings from The 14th Scandinavian Conference on Health Informatics 2016, Gothenburg, Sweden, April 6-7 2016, Linköping University Electronic Press, 2016, pp. 1-5.

[8] M. Rigby, S. Koch, D. I. Keeling, and P. Hill, "Developing a New Understanding of Enabling Health and Wellbeing in Europe: Harmonising Health and Social Care Delivery and Informatics Support to Ensure Holistic Care," European Science Foundation 2013.

[9] A. Budrionis and J. G. Bellika, "The Learning Healthcare System: Where Are We Now? A Systematic Review," Journal of biomedical informatics, vol. 64, pp. 87-92, 2016.

[10] V. Stroetman, D. Kalra, P. Lewalle, A. Rector, J. Rodrigues, K. Stroetman, G. Surjan, B. Ustun, M. Virtanen, and P. Zanstra, "Semantic Interoperability for Better Health and Safer Healthcare: Deployment and Research Roadmap for Europe," European Comission 2009

[11] Epsos: Home N.D. Available: http://epsos.eu/ November 2014

[12] D. Blumenthal and M. Tavenner "The "Meaningful Use" Regulation for Electronic Health Records," New England Journal of Medicine, vol. 363, pp. 501-504, 2010.

[13] Mission and Goals. Available: www.opencimi.org October 2015

[14] G. Fitzpatrick and G. Ellingsen, "A Review of 25 Years of Cscw Research in Healthcare: Contributions, Challenges and Future Agendas," Computer Supported Cooperative Work (CSCW), vol. 22, pp. 609-665, 2013.

[15] S. Timmermans and M. Berg, "Standardization in Action: Achieving Local Universality through Medical Protocols," Social studies of science, vol. 27, pp. 273-305, 1997.

[16] S. Van de Velde, P. Roshanov, T. Kortteisto, I. Kunnamo, B. Aertgeerts, P. O. Vandvik, and S. Flottorp, "Tailoring Implementation Strategies for Evidence-Based Recommendations Using Computerised Clinical Decision Support Systems: Protocol for the Development of the Guides Tools," Implementation Science, vol. 11, p. 29, 2016.

[17] Stortingsmelding 9, "En Innbygger, En Journal," Helse- og Omsorgsdepatementet, Available: http://www.regjeringen.no/nb/dep/hod/dok/regpubl/stmeld/2012-2013/meld-st-9-20122013.html?id=708609 November 2014.

[18] Nasjonal IKT, "Tiltak 48 (2012). Klinisk Dokumentasjon for Oversikt Og Læring," ed, 2009.

[19] H. Lærum, S. L. Bakke, R. Pedersen, and J. T. Valand, "An Update on Openehr Archetypes in Norway: Response to Article Christensen B & Ellingsen G:" Evaluating Model-Driven Development for Large-Scale Ehrs through the Openehr Approach" Ijmi May 2016, Volume 89, Pages 43-54," International journal of medical informatics, vol. 93, p. 1, 2016.

[20] A. Moreno-Conde, D. Moner, W. D. da Cruz, M. R. Santos, J. A. Maldonado, M. Robles, and D. Kalra, "Clinical Information Modeling Processes for Semantic Interoperability of Electronic Health Records: Systematic Review and Inductive Analysis," Journal of the American Medical Informatics Association, vol. 22, pp. 925-934, 2015.

[21] L. Marco-Ruiz, D. Moner, J. A. Maldonado, N. Kolstrup, and J. G. Bellika, "Archetype-Based Data Warehouse Environment to Enable the Reuse of Electronic Health Record Data," International journal of medical informatics, vol. 84, pp. 702-714, 2015.

[22] J. Pathak, K. R. Bailey, C. E. Beebe, S. Bethard, D. S. Carrell, P. J. Chen, D. Dligach, C. M. Endle, L. A. Hart, and P. J. Haug, "Normalization and Standardization of Electronic Health Records for High-Throughput Phenotyping: The Sharpn Consortium," Journal of the American Medical Informatics Association, vol. 20, pp. e341-e348, 2013.

[23] L. Marco-Ruiz, J. A. Maldonado, R. Karlsen, and J. G. Bellika, "Multidisciplinary Modelling of Symptoms and Signs with Archetypes and Snomed-Ct for Clinical Decision Support," Medical Informatics Europe - MIE 2015, 2015, pp. 125-129.

[24] D. Boscá, L. Marco, V. Burriel, T. Jaijo, J. M. Millán, A. M. Levin, O. Pastor, M. Robles, and J. A. Maldonado, "Genetic Testing Information Standardization in Hl7 Cda and Iso13606," MedInfo, 2013, pp. 338-342.

[25] R. Chen, P. Georgii-Hemming, and H. Åhlfeldt, "Representing a Chemotherapy Guideline Using Openehr and Rules," The XXII International Congress of the European Federation for Medical Informatics (MIE 2009), IOS Press, 2009, pp. 653-657.

[26] M. A. Hailemichael, L. Marco-Ruiz, and J. G. Bellika, "Privacy-Preserving Statistical Query and Processing on Distributed Openehr Data," Studies in health technology and informatics, vol. 210, pp. 766-770, 2015.

[27] S. Garde, R. Chen, H. Leslie, T. Beale, I. McNicoll, and S. Heard, "Archetype-Based Knowledge Management for Semantic Interoperability of Electronic Health Records," The XXII International Congress of the European Federation for Medical Informatics (MIE 2009), IOS Press, 2009, pp. 1007-1011.

[28] Ocean Informatics. Clininical Knowledge Manager. Openehr Clinical Knowledge Manager N.D. Available: http://openehr.org/ckm April 2014

[29] Ocean Informatics. Nasjonal Ikt N.D. Available: http://78.47.196.39/ckm/# April 2014

[30] Nehta. Nehta Clinical Knowledge Manager N.D. Available: http://www.digitalhealth.gov.au/news-and-events/news/1057-nehta-s-clinical-knowledge-manager-ckm-featured-in-health-it-to-lead-or-be-led-seminars November 2013

[31] Stuba. Ezdravje Clinical Knowledge Manager N.D. Available: http://ukz.ezdrav.si/ckm/OKM.html November 2013

[32] B. Christensen and G. Ellingsen, "Evaluating Model-Driven Development for Large-Scale Ehrs through the Openehr Approach," International journal of medical informatics, vol. 89, pp. 43-54, 2016.

[33] T. Beale and S. Heard, "Openehr Architecture Overview. Openehr Foundation," London, UK, 2008.

[34] G. C. Bowker and S. L. Star, Sorting Things Out: Classification and Its Consequences: MIT press, 2000.

[35] SNOMED-CT. Ihtsdo, Snomed-Ct®, the Organization. Available: http://www.ihtsdo.org/snomed-ct/

[36] D. Markwell, L. Sato, and E. Cheetham, "Representing Clinical Information Using Snomed Clinical Terms with Different Structural Information Models," KR-MED, 2008, pp. 72-79.

[37] Klinisk Informasjonsmodellering. Available: wiki.arketyper.no January 2016 (in Norwegian).

[38] N. R. Hardiker, D. Hoy, and A. Casey, "Standards for Nursing Terminology," Journal of the American Medical Informatics Association, vol. 7, pp. 523-528, 2000.

[39] L. Marco-Ruiz and R. Pedersen, "Challenges in Archetypes Terminology Binding Using Snomed-Ct Compositional Grammar: The Norwegian Patient Summary Case," 16th World Congress on Medical and Health Informatics (MedInfo2017), Hangzhou, China, IOS Press, 2017, p. In Press.

[40] Technical Implementation Guide - Technical Implementation Guide - Snomed Confluence N.D. Available: https://confluence.ihtsdotools.org/display/DOCTIG March 2017

[41] Hl7 Terminfo Project N.D. Available: http://wiki.hl7.org/index.php?title=TermInfo_Project# July 2017

[42] A. Rector and D. Sottara, Chap. 20 "Formal Representations and Semantic Web Technologies," in *Clinical Decision Support: The Road to Broad Adoption*, R. A. Greenes, Second Edition. Oxford: Academic Press, 2014, pp. 551-598.

[43] Clinical Models and Snomed Kaiser Perspective - Cimi N.D. Available: http://informatics.mayo.edu/CIMI/index.php/Clinical_Models_and_SNOMED_Kaiser_Perspective July 2017

[44] L. Marco-Ruiz, C. Pedrinaci, J. A. Maldonado, L. Panziera, R. Chen, and J. G. Bellika, "Publication, Discovery and Interoperability of Clinical Decision Support Systems: A Linked Data Approach," Journal of Biomedical Informatics, vol. 62, pp. 243-264, 2016/08/01/ 2016.

[45] R. Pedersen, G. Ellingsen, and E. Monteiro, "The Standardized Nurse: Mission Impossible?," Researching the future in information systems, pp. 163-178, 2011.

[46] Helse Nord Fiks – Én Journal I Nord. Available: https://helse-nord.no/helse-nord-fiks-n-journal-i-nord January 2016 (in Norwegian).

[47] C. Martínez-Costa, R. Cornet, D. Karlsson, S. Schulz, and D. Kalra, "Semantic Enrichment of Clinical Models Towards Semantic Interoperability. The Heart Failure Summary Use Case," Journal of the American Medical Informatics Association, vol. 22, pp. 565-576, 2015.

[48] J. E. Huerta, C. A. Villar, M. C. Fernández, G. M. Cuenca, and I. A. Acebedo, "Nhs Electronic Health Record," Ministerio de Sanidad y Politica Social, Available: http://www.msssi.gob.es/organizacion/sns/planCalidadSNS/docs/HCDSNS_English.pdf March 2014.

[49] Health Level Seven. Hl7 Standards Product Brief - Cda® Release 2. Available: http://www.hl7.org/implement/standards/product_brief.cfm?product_id=7 March 2015

Evidence-Based Self-Management for Spondyloarthritis Patients

Wolfgang Leister, Ingvar Tjøstheim,
Trenton Schulz, Peder Aursand, Aleksander Bai
Norsk Regnesentral
Oslo, Norway
email: {wolfgang.leister, ingvar.tjostheim,
trenton.schulz, peder.aursand, aleksander.bai}@nr.no

Kåre Birger Hagen, Camilla Fongen,
Nina Østerås, Hanne Dagfinrud
NKRR, Diakonhjemmet Hospital
Oslo, Norway
email: k.b.hagen@medisin.uio.no
camilla.fongen@diakonsyk.no

Lars Thomas Boye
Tellu AS
Asker, Norway

Vebjørn Berre
APERTUS AS
Asker, Norway

Liming Chen
De Montfort University
Leicester, UK
email: liming.chen@dmu.ac.uk

Dag Magnar Soldal, Carl-Henrik Franke
Hospital of Southern Norway
Kristiansand, Norway
email: {dag.magnar.soldal,
carl-henrik.franke}@sshf.no

Jon Hagfors,
Jan Halvard Relbe-Moe
Norsk Revmatikerforbund
Norway

***Abstract*—We present a concept including a set of tools for self-management for patients suffering from axial spondyloarthritis (SpA). This concept involves patient-recorded outcome measures, both subjective assessment and clinical measurements, that are used to present recommendations. We report from experiences made while implementing a proof of this concept and analyse it from several perspectives. Our work resulted in proposing a self-management tool for the patient, improving the methodology for clinical measurements of rotation exercises, and proof the viability for using on-board sensors in smart phones. Further, since sensors collect data in a medical setting, we present ethical considerations.**

Keywords*—*axial spondyloarthritis; self-management; health care; self-assessment; evidence-based; mobile applications; sensors; ethics.

I. Introduction

In our previous work [1], we presented a concept for evidence-based self-management of patients suffering from axial SpA. We have updated this concept and extended the implementation of a smart phone and sensor-based system that can give recommendations to the patients as support for self-managing their condition.

For a variety of chronic diseases, patients managing the condition themselves (*self-management*) can result in reduced costs in the health care sector and an improved clinical outcome [2]. Self-management encompasses methods where the patients participate in managing their disease through education and changes in behaviour and lifestyle [3] [4]. In evidence-based self-management, elements such as clinical assessment, collaborative priority and goal-setting, patient self-efficacy, and active follow-up are essential [5]. We look closer at self-management settings [6] where patients assess the status of their disease using sensors and questionnaires on their smart phones and report the results (i.e., patient-reported outcome measures) [7] [8]. Based on these measurements and self-reported outcomes patients receive non-pharmacological recommendations from the self-management system to increase their coping skills, help with pain management, adhere to their medication regime, improve self-care behaviours, and enact lifestyle changes.

Spondyloarthritis (SpA) describes a group of several related, but phenotypically distinct rheumatic diseases, such as ankylosing spondylitis (AS). The condition *axial SpA* is characterised by inflammatory back pain and mainly affects the axial skeleton, which is distinct from peripheral SpA where the symptoms are mainly arthritis, entesitis, or dactylitis. In axial SpA, the first appearance is mainly in young adulthood and can lead to structural and functional impairments and a decrease in health related quality of life. Although axial SpA is a chronic condition, the symptoms and disease activity vary over time [9] [10]. The primary goals for managing axial SpA are to maximise long term health-related quality of life by managing symptoms and inflammation, preventing progressive structural damage in the spine, and normalising function and social participation. Relevant medication and non-pharmacological treatment such as physical training are recommended as the foundation of the management of axial SpA [11] [12].

Currently, there are few evidence-based self-management tools for axial SpA. Some tools for subjective assessment exist, but sensor-based tools for objective assessment are not yet available to the wider public. Also, there are obstacles to let patient-assessed data be of use in a clinical setting [13].

This paper extends our concept for evidence-based self-management of axial SpA. This concept is supported by an implementation of a smart phone and sensor-based system that can give recommendations to the patients. We report from experiences from this implementation. We also perform an ethics assessment and risk analysis of our concept.

The remainder of this paper is organised as follows: After a brief presentation of related work (Section II) and presenting our concept of self-management for axial SpA patients (Section III), we show details from the proof of concept implementation involving subjective and sensor-based clinical assessment and recommendations to the patient (Section IV).

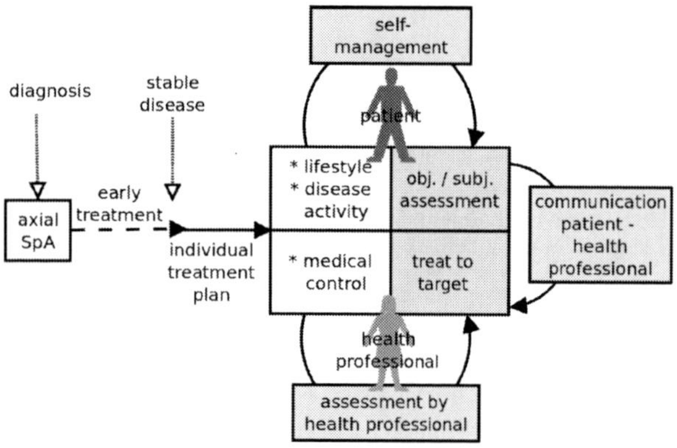

Fig. 1: Concept for managing the disease with the three parts: self-management, clinical assessment, and patient-health personnel communication.

Then, we present the results from a usability test (Section V). Further, we look into regulatory issues, ethical challenges, and perform an informal risk analysis that identifies functionality that needs to be implemented before productification of our concept can be done (Section VI). Finally, we discuss our findings (Section VII) and conclude in Section VIII.

II. Related Work

For the management of most chronic illnesses the patients and their carers have an extensive responsibility regarding adherence to the treatment plan, life style changes, taking preventive actions, etc. Newman et al. [14] presented a literature review of self-management interventions for chronic illnesses, here under arthritis, asthma, and diabetes. They reported about the content of interventions as well as outcomes. Note that since Newman et al.'s review in 2004 new concepts for self-management have been developed, specifically those that make use of emerging technologies, such as smart phones, sensors, and actuators.

A. Self-Management

We find multiple definitions of *self-management* in the literature. The term self-management covers all means of empowering individuals to cope with disease and experience a high quality of life by developing self-efficacy. Self-management also refers to an individual's ability to manage the symptoms, treatment, physical, psychosocial, and lifestyle changes inherent in living with a chronic condition [15]. In this context, *self-efficacy* is the individual's level of confidence in succeeding to cope with that individual's chronic disease.

Intervention elements used in self-management often include education, follow-up strategies, motivational counseling, and individualised care plans. Johnston et al. [15] presented a literature review showing success factors and limitations of self-management.

Considerable work has been done on self-management programs for chronic diseases with good results in terms of quality of life, and reducing the need for care and cost efficiency [16]. Programs such as *The Chronic Disease Self-Management Program* have shown significant improvement in health distress and increased perceived self-efficacy [17]. The motivation for these programs is to provide people with chronic diseases the tool to efficiently manage their own condition.

We prefer the definition of self-management by Barlow et al. [18] "…the individual's ability to manage the symptoms, treatment, physical and psychosocial consequences and life style changes inherent in living with a chronic condition." Barlow et al. stressed that monitoring one's condition and the effect of responses to daily life can lead to a dynamic and continuous process of self-regulation.

B. Mobile eHealth Apps

The number of mobile health (mHealth) apps available to consumers exceeded 165 000 in 2015 [19] and is still growing. According to this report, two-thirds of these apps are related to wellness management, i.e., fitness, lifestyle, diet, and nutrition. Further, a quarter of these apps are related to disease and treatment management; about 9% of the total number of apps is disease-specific. About 6-7% of the disease-specific apps relate to musculo-skeletal diseases, where SpA is placed.

Dimensions of app-functionality include *a*) to inform, *b*) to instruct, *c*) to record, *d*) to display, *e*) to guide, *f*) to remind, and *g*) to communicate [19]. For evidence-based concepts, aspects *c* to *g* are of most relevance.

We have found management support for some chronic conditions using information and communication technology (ICT). These include: a self-management application called SoberDiary for alcoholism [20], a programme for quitting smoking [21], a programme for psychology support [22], a mobile application for diabetes that integrates with personal smart watches [23], a virtual coach for chronically ill elderly [24], a smart phone app for rheumatic diseases management [25]. There are also generic apps for integrating vital signs into personal health devices or electronic medical record systems [26].

A review of the effectiveness of multiple European eHealth initiatives concluded that while eHealth can be very effective, certain criteria need to be fulfilled for the success of such interventions [27].

Within axial SpA there are ICT apps like *SpA Helper* [28] that supports to monitor the disease. When *SpA Helper* is used, the results from the monitoring are not part of a feedback cycle, such as the treat-to-target principle (see Section III-A). There are also a variety of web-based calculators that implement the subjective clinical indices.

C. Sensors

There is extensive research on using sensors for tracking physical activity that ranges from physiological wearable sensors to the using mobile phones for tracking activities [29]. In recent years, we observed an increase in the number of dedicated activity trackers like FitBit Charge or Garmin Vivosmart. Research and user-testing have been conducted to monitor physical activity in a non-obtrusive manner, but more research is needed to investigate which methods and devices work best for different demographics [30].

In general, sensors can only measure physical or physiological data that represent the patient's physiological reactions or data from the environment, i.e., context information about the user. Subjective data still need be retrieved using questionnaires or similar methods. Note that technologies such as brain-computer interfaces (BCI) [31] are too obtrusive. Further, the BCI technology is still in an early state.

Often sensors measuring factors that can determine emotions (such as cameras, sound analysis, or skin conductance in the research field of *affective computing* [32]) can be used for well-being applications. Such technologies have been used to assess emotions of visitors in science centers [33]. However, for the purpose of assessing a single person's state, there is only a statistical correlation between mood recognised by an algorithm and a person's real state. Thus, such measurements cannot be used as input for medical decisions in the treatment of axial SpA.

III. A System Architecture for Self-Management

Our concept is based on medical principles that are applied to a computer-based system with several components: sensors, smart phone, a health cloud, patient records and communication between these components. The system architecture addresses medical principles, restrictions caused by technical reasons, and policies by the stakeholders.

A. Integrating Treat-To-Target

A *treat-to-target method* [34] has been developed for treating axial SpA. This evidence-based method is used after diagnosis and early treatment, when the disease has reached a stable state (Fig. 1). At this stage, an individual treatment plan has been created for the patient. This method uses a treatment goal (*target*) for a *treatment plan*. Following the treatment plan and regularly assessing the patient's status provides evidence about how the disease develops. When the patient's status moves away from the treatment target to a worse condition, health personnel, in discussion with the patient, might adjust the treatment plan or target.

As part of an evidence-based, self-management setting, the treat-to-target method is extended so the patient can perform self-assessment to gather evidence about the current disease condition by performing assessments, answering questionnaires, and following the progress from the patient diary. The patient diary data can be used for patient-health personnel communication by making it available to the clinical personnel, either regularly or when needed (e.g., a patient visit).

Fig. 2 shows how treat-to-target can be aligned with self-management. The upper unshaded part of the drawing is the health personnel domain. This is where health personnel perform clinical assessments and decide the treatment target and treatment plan. The lower shaded part is the patient's domain. This is where the patient can perform assessments, compare with the target, and adjust some elements of the treatment.

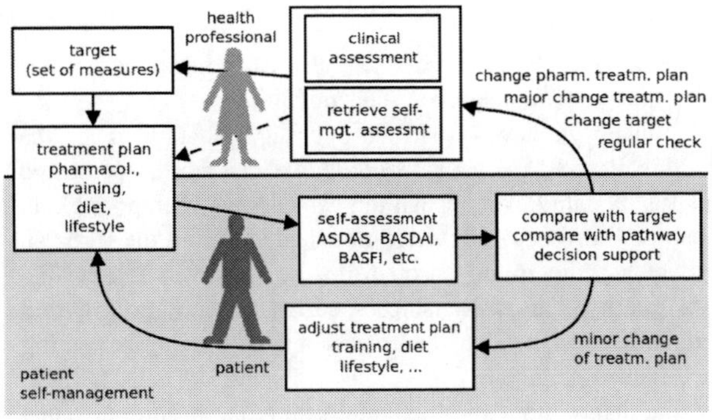

Fig. 2: Treat-to-target in a self-management setting, showing tasks to be performed by patient and health personnel, respectively.

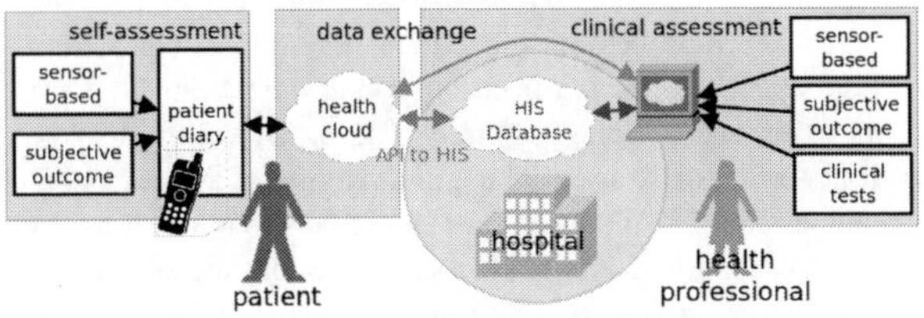

Fig. 3: Architecture of a self-management system including three parts: self-management, clinical assessment, and data exchange.

B. An Architecture for axial SpA Treatment

Our concept (Fig. 3) builds on *a*) a solution for self-management, *b*) better quality and effectiveness of *clinical assessment*, and *c*) enhanced *patient-health personnel communication*.

The solution for self-management lets the patients use tools at home to manage the disease. It includes patient-reported outcome measures [8], the assessment of ample parameters, the use of a patient diary [35], patient guidance with respect to the treat-to-target principles, and alerts in case of changes in the patient's condition or physical function.

The concept also enhances the quality and effectiveness of clinical assessment; assessment methods developed for self-management are made available for clinical assessment.

The concept includes a foundation for patient-health personnel communication. Self-reported assessments can be used for patient-health personnel communication to explain or visualise the development of the disease and data transfer to the hospital.

C. The Health Cloud

Our concept uses a health cloud to facilitate persistent storage, and as a means to communicate data to the health care personnel. Although this is an extra component, there are several reasons for this health cloud: *a*) persistence and storage of health data and keep data consistent over several devices the patient might use; *b*) give the patient ownership over the patient data and the possibility to structure these after the patient's own decisions; *c*) give the patient the possibility to exchange clean data to the health authority's system (e.g., patient records) at the patient's discretion; and *d*) give the patient the possibility to store data also when health authorities have not (yet) implemented such possibilities.

Although we have considered exchanging data between the health cloud and the patient records, we have not implemented this functionality even though this would have been possible in Norway. Extra costs that were not funded during the research project are the main reason for this. However, issues like data quality, data ownership, security, privacy, policy, and standardisation need to be considered before implementing such data exchange.

Note that without such communication, the health personnel can get access to the patient data either by accessing the data on the patient's smart phone or by accessing the health cloud via a specifically authorised interface. We also note that introducing a further interface beyond the patient records will probably not be a success criterion in the introduction of our concept. To our experience, health personnel will not be willing to take further interfaces into use. Colleagues of some of the authors at Diakonhjemmet Hospital and the Hospital of Southern Norway were interviewed, and they made it clear that introducing further special purpose interfaces would distract them from their daily work. Thus, implementing the API between the health cloud and the health records will be the only long-term option.

D. Development Methodology

The concept and implementation of our prototype was performed interdisciplinarily as an innovation project in the public sector, where major parts of the health care sector in Norway belong. The development team included health care personnel specialising in rheumatology and physiotherapy at hospitals. They could also draw on the expertise of their colleagues. The development team also included two representatives from the Norwegian Rheumatology Association to give the team a user-centred design focus. Further, computer scientists, programmers, and developers were involved in the work that also included technical and medical evaluations.

IV. Proof of Concept for axial SpA Self-Management

The parts of this architecture that include data exchange between a health cloud or a patient's devices and the electronic health record (EHR) system are beyond the scope of our work. These are parts that rely on policies defined by public health care providers. So, we focused on implementing tools for clinical assessment and self-management.

A. Medical Assessment Methods for axial SpA

Medical self-assessment is essential for evidence-based self-management. So, these self-assessment methods should be based on medical assessment methods since evidence for their effectiveness is documented.

The *AS Disease Activity Score* (ASDAS) is used for measuring and monitoring disease activity in axial SpA. It is based on a composite score of domains relevant to patients and clinicians, including both self-reported items and objective measures [36].

The Bath indices [37] present outcome measures for use with axial SpA patients, and consist of four indices: the Bath AS Metrology Index (BASMI), the Bath AS Functional Index (BASFI), the Bath AS Disease Activity Index (BASDAI), and the Bath AS Patient Global Score (BAS-G). These indices are designed to give a good clinical assessment using a minimum number of measurements or questions to be answered. The BASMI is five simple clinical measurements; the other indices consist of a number of questions that are answered on a numeric scale from zero to ten [38].

Østerås et al. [39] described a set of assessment tests as candidates for axial SpA self-assessment. These exercises are: lateral spinal flexion, modified Schober's, cervical rotation, occiput to wall distance, tragus to wall distance, intermalleolar rotation around the vertical axis, lumbal and thoracic rotation, six-minutes walking test, stair climb test, sit-to-stand test, fingertip-to-floor test, and maximum grip strength test.

B. Requirements for Sensor-based Self-assessment

Clinical measurements for self-assessment must, ideally, be implemented in a way that allows the patient to operate the measurement at home and without assistance. As patients with axial SpA have or may develop a restricted range of motion, the equipment needs to be designed so that it is possible for the target group to use and setup any devices without assistance.

Further, the measurements must be designed so that incorrect operation by the patient is unlikely when performing the exercises and measuring. If possible, deviations should be recognised by the system, indicated to the patient, and transferred measurement results should be marked as invalid.

To implement these measurement processes, we can use a variety of sensor types, such as inertial sensors for movements, magnetic sensors, cameras, and microphones. Sensors built into, e.g., smart phones or smart-watches, can be used for such measurements if this is practical. External devices, often connected via a wireless medium, e.g., Bluetooth, can be better adapted to certain types of measurements and, thus, deliver more precise results. Further, there are specific external devices for measurements where suitable sensors are unavailable in standard smart phones, e.g., sensors for diabetes. For some exercises, training equipment with built-in sensors (spinning machine) could be used. Most specific sensors could be implemented with connectivity to a smart phone (e.g., communication protocols using Bluetooth, ANT+, or similar).

For some types of measurements, the built-in sensors of smart phones could be used, e.g., the measurement of the heart rate of a person using the LED-flash and camera of a smart phone [40]. Further, inertial and magnetic sensors, camera, and microphone could be used to implement measurements [41] [42]. There is a variety of training-related apps that make use of both internal sensors as well as external sensors and training armbands. As a note, the use of GPS-positioning seems less useful for self-management for axial SpA patients.

Consumer devices that are currently available include smart-watches and bracelets. Some of these are programmable and have ample sensors on-board. Such devices are unobtrusive and can be used in most daily-life situations.

We foresee that new unobtrusive devices and sensors will be developed and integrated into textiles and spectacles that can be worn by the person. Alternatively, they may even implanted into the person. But with these new devices arise ethical problems that could violate the rights of the person or third parties. Thus, suitable counter-measures need to be taken to avoid this.

C. Implementing Exercises for Self-Assessment

Considering the exercises to assess axial SpA conditions, we can consider classes of exercises where the form factor needs to be adapted for the specific exercise.

1) Limb Movements and Position: Exercises that assess the relative position of limbs after a movement (e.g., the spinal lateral flexion or the fingertip-to-floor test) can be implemented using an inertial sensor. For the assessment, the sensor is attached to the limb, the position is assessed in the reference position, the movement is performed, and the position after the movement is assessed. The distance between these positions is reported. Challenges for such assessments is how to attach the sensors and the user interface (how to trigger the assessment).

Using a camera and employing image processing algorithms would be another solution [43]. Zhao et al. [44] presented such a system to track health workers' positions while carrying out their work with patients. However, placement of the camera, lighting conditions, and physiological conditions (body shape) can result in challenges and reduced accuracy.

2) Rotation Exercises: Exercises that include circular movements of body or limbs (e.g., cervical, lumbal, thoracic movements; intermalleolar rotation around the vertical axis) can be implemented using an inertial sensor and, to a certain degree, a magnetic sensor. For the assessment, the sensor is attached to the body part to be rotated, a reference position is assessed, the movements are performed, and, at each of the extreme points, the relative movement is assessed.

3) Exercises that Need Specific Devices: For several of the exercises the assessment requires either specific sensors or external assistance. The modified Schober's assessment measures differences in the length of the spine. This measurement cannot be performed without the assistance of a physiotherapist. However, the fingertip-to-floor test is correlated with the modified Schober's test [45].

Assessment of the distances of the tragus or occiput to the wall could be implemented using some specific device, e.g., using a ruler; this, however is beyond the scope of our project. Further, the assessor needs to ensure that both heels and back touch the wall for a correct assessment [38].

The maximum grip strength test is used to determine muscle strength. Currently, it can only be implemented using specific sensors or devices such as a dynamometer.

4) Tests for Health-Related Physical Fitness: Exercises that assess health-related physical fitness seem at first glance rather easy to implement. But as we shall see, there are several issues with these tests.

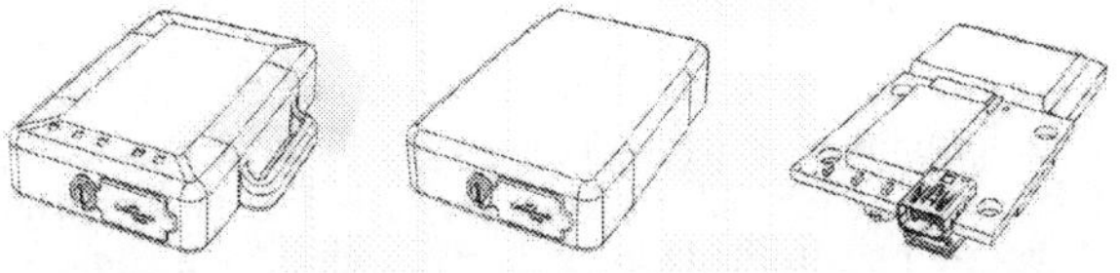

Fig. 4: Drawing of the APERTUS sensor used for axial movements.

For the sit-to-stand test (also called the 30-second Chair Stand Test or 30s CST), the patient is given 30 seconds to complete as many full stands as possible starting from a seated position with arms folded across the chest. Such assessments can use the same principle as the limb movements, counting the number of up- and down movements. However, with this method it is not evident whether these movements have been only partially performed. We note that this test could be used for self-management due to its simplicity.

The six-minutes walking test (6MWT) is an inexpensive and simple walking test that can be used as a predictor of aerobic fitness or for assessing the sub-maximal level of functional capacity. Following the American Thoracic Society guidelines [46], patients are instructed to walk as fast as possible back and forth between two cones 15 meter apart on a flat, hard surface for six minutes. The walking distance is measured in meters. During the test, the heart rate is recorded with a heart rate monitor, and perceived exertion can be measured with Borg's rating of perceived exertion (RPE) [47].

The test is claimed to be most useful in patients with at least moderately severe impairments while other more well-functioning populations may exhibit a ceiling effect since the test is limited to a patient's walking speed.

While there is a variety of sensors that could be used to implement an assessment, including indoor positioning, inertial sensors, step counters, the most important question is whether the necessary space to perform such an exercise is available in the patient's home. If available, e.g., some patients might have access to a corridor to perform the exercise, patients might not want to use this in a space that is not private.

In the stair climb test (ST) [48], the patient is asked to ascend and descend 50 normally sized steps with three landings in-between as quickly as possible, also measuring the heart rate. This test could be implemented using step counters or intertial sensors. Note that not all patients might have access to suitable stair steps.

D. Sensor-based Clinical Measurements

APERTUS developed a sensor that can measure rotation around the vertical axis such as cervical rotation, thoracic rotation and hip abduction (measured in the supine position). The Apertus sensor was developed by engineers in close collaboration with health professionals and patients with axial SpA in a process with discussions, development of prototypes and (re-)testing.

The result can be transmitted via a wireless connection to a receiver, such as PC, tablet, or smart phone. This inertial sensor is packaged in a small box (Fig. 4) that can be attached to the body. The size of the device is $55\text{mm} \times 35\text{mm} \times 3\text{mm}$. The sensor contains radio technology that follows Bluetooth standards that might influence electronic devices in 2.4 GHz ISM, but to a significantly lower degree than mobile phones.

Compared with other technology such as lasers or optical sensors, this sensor's advantages include its high precision and being cheaper, smaller, and lighter than the other solutions. Compared to the traditional way of measuring rotation with a goniometer or myrinometer (e.g., compass) the sensor provides more precise measurements. The sensor is a simple way to achieve satisfactory measurements in acceptable use of time and without health personnel assisting.

The Apertus sensor was tested in a laboratory setting and the data collected was compared with Cybex 6000 simulation data as gold standard. The sensor was mounted to the cybex with a bracket in six different positions, *1)* straight up; *2)* straight down; *3)* up, tilted 30° around the frontal axis; *4)* down, tilted 30° around the frontal axis; *5)* up, tilted 30° around the frontal axis, 20° around the sagittal axis; and *6)* down, tilted 30° around the frontal axis, 20° around the sagittal axis. Reference angles were set at approximately 10, 30, 60, 90 and 120 degrees. From a center position the sensor was moved from left to right (one cycle) by manual force at the speed allowed ($60°\,s^{-1}$) until stop brackets were reached. Criterion validity and reliability for single measures and for the mean of three trials (left, right or cycle) was evaluated with two ways mixed interclass correlation coefficient (ICC). Limits of agreement (LoA) (Bland and Altman method), and smallest detectable change (SDC95%) with 95% CI were calculated to evaluate the measurement error of the sensor. The sensor showed excellent criterion validity and reliability for rotation around the vertical axis in the range of motion from 10° to 120°. The angle can be measured with a precision of ±0.87 for one measure and 0.80 for the mean of three measures. When assessing a single cycle or the average of three cycles, a change of 1 degree is needed for real change beyond measurement error. These findings justify proceeding with further evaluations of the sensor for this kind of measurements [49] [50]. A clinical trial of the rotation measurements with 60 patients suffering from axial SpA is currently under evaluation.

We developed a suitable user interface for the assessment process with the sensor. The assessed data are stored locally in the patient diary and forwarded to the health cloud for permanent storage.

E. Smartphone-based Self-Management Measurements

Self-management based on smart phone platforms carries some benefits over dedicated third-party measuring devices. The user is likely to be more familiar interacting with a personal phone than any new measuring instruments. Moreover, the convenience of using a device already carried may significantly increase user acceptance and on-boarding. Further, by taking advantage of accurate high quality built-in sensors already available on smart phone platforms, the extra cost of acquiring measuring instruments can be avoided.

(a) Measurement screen. (b) History plot screen (c) History data screen.

Fig. 5: Screen shots from the MOSKUSapp proof of concept for rotational exercises.

As a proof of concept, an Android app has been developed that allows the user to measure cervical rotation, thoracic rotation and hip abduction using the phone's internal gyroscope. When starting a new measurement the phone can be positioned on the forehead (using, e.g., a regular head band) or in a pocket on the torso or on the hip. When activated, the app gives sound queues at regular intervals indicating when the user should turn forward, turn left, turn forward, turn right, and turn forward again. The sound queues allow for the use of the app during rotational exercises when the screen is not visible to the user. At the end of the measurement the user is presented with the results and given an option to save the data, allowing tracking of trends over time. The basic elements of the user interface are shown in Figure 5.

Low sensor accuracy and lack of uniformity represent potential drawbacks of the smart phone approach when compared to the use of dedicated sensors. A gyroscope sensor has been available on the Android platform since version 2.3 (API 9). In contrast to the iPhone, where hardware is largely homogeneous and standardised, the quality of sensors on an Android phone can differ greatly between manufacturers. However, some universal requirements exist for licensed Android smart phones. As of 2016, gyroscopes on licensed Android 7.0 devices are required to be calibrated, persistent between reboots, temperature compensated, have a resolution of at least 12 bits and a variance of no greater than $1 \times 10^{-7}\,\text{rad}^2\,\text{s}^{-1}$ (or approximately $0.02\,\text{deg}/\sqrt{\text{s}}$). However, all gyroscopes will exhibit a drift over time, and on low-end devices this drift can potentially influence angle measurements.

To quantify the potential inaccuracy when using built-in smart phone gyroscopes for the measurement of cervical rotation, a simple experimental procedure was carried out. The prototype app was installed on four different smart phones of varying age: Sony D5803, Sony D6603, Samsung S3 and Samsung S8. The newest phone in the test was the Samsung S8 (released April 2017) and the oldest was the Samsumg S3 (released May 2012). The phones were rotated 30°, 60°, and 90° at a radius of 0.1 m with a rotation time from the neutral

TABLE I: The bias and standard deviation of the measurement error in degrees using the proof of concept Android app with gyroscope for measuring cervical rotation of 30°, 60°, and 90° with different phones.

	30°		60°		90°	
Phone	$\bar{\Theta}$	STD	$\bar{\Theta}$	STD	$\bar{\Theta}$	STD
Sony D5803	-0.33	0.68	-0.72	0.96	-0.74	0.68
Sony D6603	-0.60	0.81	-0.35	0.82	0.09	0.61
Samsung S3	-1.01	1.08	-1.09	1.22	-2.50	1.48
Samsung S8	0.12	0.92	-0.15	0.84	0.23	0.77
Overall	-0.65	0.86	-0.72	1.00	-1.05	0.92

TABLE II: The bias and standard deviation of the measurement error in degrees using the proof of concept Android app with accelerometer for measuring cervical rotation of 30°, 60° and 90° with different phones.

	30°		60°		90°	
Phone	$\bar{\Theta}$	STD	$\bar{\Theta}$	STD	$\bar{\Theta}$	STD
Sony D5803	15.0	8.59	10.45	19.57	9.51	20.71
Sony D6603	23.98	9.10	42.05	8.94	107.7	32.53
Overall	19.5	8.84	26.25	14.26	58.61	26.62

position of about 1 s. The rotation was alternated between left and right and the experiment was repeated 10 times per device. In each case the difference in the angle measured by the app, Θ_i, and the known angle of rotation, Θ_0, was recorded. Table I shows the bias

$$\bar{\Theta} = \frac{1}{N}\sum_{i=1}^{N}(\Theta_i - \Theta_0) \quad (1)$$

and the standard deviation

$$\text{STD} = \sqrt{\sum_{i=1}^{N}(\Theta_i - \bar{\Theta})^2} \quad (2)$$

in degrees when using the different Android devices.

The results shown in Table I indicate that the overall measure error when using smart phone built-in gyroscope is of a comparable magnitude as the error reported in the sensor-based clinical measurements. Overall, the error was measured to be 0.61° to 1.48°. As expected, the error (STD) is somewhat larger for greater rotation angles and for the older of the phones tested (Samsung S3).

We observed also that some devices seem to produce useful results for angles up to 90°, but show large deviations for angles above. This is one of the reasons why we changed the measurement procedure for the rotation exercises to measure maximum 90°, rather than up to 180° as done in our first implementation.

While more rigorous testing is required before such a mobile application can be put into use, our tests indicate a clear potential of using smart phone built-in gyroscopes for measurement of cervical rotation.

Some (older or less sophisticated) smart phones come without a built-in gyroscope. For such cases one can consider using the phone's accelerometer to estimate the arc length when rotating the phone, and calculating the rotation angle using an assumed radius. The angular inaccuracy for this approach was estimated using two different smart phones and the same experimental setup as used previously. Table II shows that the inaccuracy caused by accumulated accelerometer noise when using such a method renders this approach unfeasible. As a note, we also tested other smart phones such as the Samsung S3 and S8, but the results were entirely unusable and, thus, omitted from the table. It should be noted that the results when using accelerometer will to some degree depend on the algorithm used to compute the angle and may be improved upon. In this test a first-order Euler integration was carried out to get the arc length when rotating. The angle was obtained assuming the 0.1 m rotation radius.

In summary, a number of the self assessment tests identified by Østerås et al. [39] can, potentially, be carried out using a smart phone platform. For example, one can envisage the measurement of the six-minutes walking test, the stair climb test and the sit-to-stand test using the phone's built-in accelerometer combined with voice instructions. Future work includes extending the proof of concept self-management Android app to also include some of these features.

F. Self-Assessment of Subjective Conditions

For the assessment of the subjective conditions for BASDAI and BASFI we implemented suitable user interfaces in our prototype (Fig. 6). We also implemented a questionnaire for ASDAS, a composite score including subjective evaluation and the inflammatory markers C-reactive protein (CRP) and erythrocyte sedimentation rate (ESR). Patients answer the questions on a scale from zero to ten by tapping on the appropriate number. We did not choose sliders because we assumed that tapping on the appropriate field would be easier for the target group with their possible movement restrictions. Note that the use of a numeric rating scale is recommended by ASAS [12].

After the form is finished, the data are stored locally. An estimate of the current health condition is shown to give the patient feedback along with the possibility to report these data to the health cloud for permanent storage.

Questionnaires can be scheduled using the mobile device's calendar by creating calendar entries with a specific syntax. The calendar then reminds patients to perform assessments at a given time.

G. Self-Management and Recommendation

Development of the self-management was based on the ASAS-EULAR management recommendation for axial SpA [12] taking into account the opinion of health professionals working within the field of rheumatology, patients with axial SpA, a psychologist and developers of the diary app.

A self-management system needs to support the patient in the following ways: *a*) deciding the type and degree of adjustments for non-pharmacological changes in a treatment plan, such as diet, training, lifestyle, or other minor adjustments; *b*) identifying significant deviations from the expected progress and present these deviations to the patient and health personnel; *c*) advising changes of treatment plan to the health personnel;

(a) The ASDAS/ BASDAI menu. (b) The BASFI menu. (c) The result page.

Fig. 6: Screenshots for the data collection module.

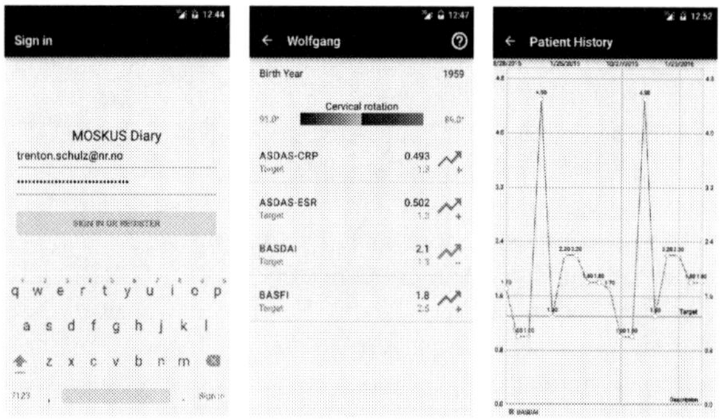

(a) The diary login screen. (b) The diary patient view. (c) The diary history view for BASDAI.

Fig. 7: Screenshots from the patient diary app.

and *d*) suggest changes of patient's target to the health personnel. Qi et al. [51] present an approach for how to make decisions that are presented to the patient. We use a diary in our solution.

The diary shows the disease's development visually, deviations from the treatment plan, and gives recommendations using *trend labels*. The patient view (Fig. 7b) shows the patient's birth year, the current left and right cervical rotation, and the current scores for ASDAS, BASDAI, and BASFI, including their targets. Each of these scores can have historical data; this is shown in the patient history view (Fig. 7c) and summarised in the trend labels (Fig. 8) to the right of the value. There are five trend labels: (a) disease activity is increasing, but below target; (b) disease activity is increasing; (c) no change in disease activity; (d) disease activity is decreasing, either below or heading towards target; and (e) disease activity is decreasing, but still very high and needs more treatment.

(a) (b) (c) (d) (e)

Fig. 8: Trend labels for the patient view.

H. Heuristic Decision Support Based on Medical Expertise

We created simple rules to guide patients and health personnel. These rules are based on the values from ASDAS, BASDAI, and BASFI.

The values derived in ASDAS and BASDAI indicate the amount of disease activity. Machado et al. [52] define cutoff values for disease activity measured using ASDAS: *a*) under 1.3 – the disease is inactive; *b*) between 1.3 and 2.1 – disease activity is moderate; *c*) between 2.1 to 3.5 – disease activity is high; and *d*) over 3.5 – disease activity is very high. A change on the ASDAS scale of 1.1 or more is considered a clinically important change while 2.0 or more is considered a major change. Based on this work, we indicate the trend of the scores (up, down, or steady), as well as the severity (colour). The thresholds can be personalised for patients where health personnel defines alternative values.

Braun et al. [53] propose a similar approach for BASDAI by calculating a trend line that uses BASDAI targets for cutoffs. Situations where the BASDAI is above 4.0 – indication of high disease activity – or changes of the BASDAI over 50% or a factor of two also generate a warning to contact health personnel.

BASFI indicates the disability level. Wariaghli et al. [54] ran a large survey with Moroccan patients and defined the target values depending on the patient's age in three age ranges. We use similar rules as above for determining the trend based on the patient's target or age information, depending on what is available.

Recently, Kviatkovsky et al. [55] proposed cutoff values for the patient-acceptable symptom state (PASS) for BASDAI and BASFI to be 4.1 and 3.8, respectively. For the minimum clinically important improvement (MCII) the values were 0.7 and 0.4, respectively. However, for patients with an active disease, the MCII values are 1.1 and 0.6, respectively.

V. USABILITY TEST OF THE PROTOTYPES

Usability, how easy something is to use, is an important factor for adoption and continuous use of a system or application. Motivated by this, we performed a usability test of the two developed prototypes. We wanted to see how appropriate the apps are for their purpose, and to get feedback on the usability. We employed the System Usability Scale (SUS) developed by Brooke [56]. The SUS consists of ten questions that are rated by the participants on a five point or a seven point Likert scale [57]. The ratings are used to calculate a score on a scale from 0 to 100 where 70 is the average score. Additionally, we added six questions on related matters that are not part of the SUS scale, e.g., the need for the apps, satisfaction, and whether participants would recommend the apps to others.

For the usability test, we recruited eighteen individuals with Android smartphones among members of the Norwegian

TABLE III: SUS scores for the collection module app and patient diary app with highest and lowest scores removed.

	Average	Median
Collection module	73.73	80.20
Patient diary	74.05	74.35

Rheumatology Association (Norsk Revmatikerforbund). We asked the eighteen to download the two apps and sign up for the usability test. Of the invited participants, fourteen followed the procedure, downloaded the apps, and registered at the health cloud site. The individuals received the link to the surveys after they had downloaded the apps and a text message with instructions. Of the eighteen individuals, nine used the apps and completed the test.

With only nine respondents, the usability test is more a pilot study. If an application has major usability weaknesses, these will likely be revealed with small sample sizes also. Our test did not indicate major weaknesses. We did not perform statistical analysis of the data beyond calculating SUS scores.

We performed separate tests for each app. The average and median from the SUS are presented in Table III with the highest and lowest scores eliminated. The SUS scores of the apps are around 74, which is just above average. As the prototype apps have several known weaknesses, this is what we can expect. Note that the app for smartphone-based measurements of rotation exercises was developed after this test has been finished; thus, the SUS scores for the measurement app are not available.

VI. Ethics, Regulatory Issues, and Risk Analysis

Using devices, sensors, and supporting health care systems may create a variety of ethical issues that need to be analysed and addressed before a system using these is used by patients. Further, there is a variety of regulatory issues that must be addressed.

The ensemble of devices, sensors, and supporting health care systems is also referred to as the *health-related Internet of Things.* (IoT) which refers to uniquely identifiable objects (things) and their virtual representations in an Internet-like structure. This term was first used in 1999 by Ashton [58]. Other definitions of IoT have appeared as technology progresses. A *thing* is a real or virtual object, e.g., a device or a web service, offering one or more services. The IoT is today a rather common platform for the deployment of health services. While our work makes use of sensors and networks, we focus on the concept for self-management rather than proposing a full deployment of our concept in the IoT.

A. Ethics Assessment

Mittelstadt [59] presents a literature review of ethics in the health-related IoT for devices, data protocols, and mediated care. These aspects can arise from both implementation and deployment. These aspects include: *i*) personal privacy; *ii*) obtrusiveness, stigma, and autonomy; *iii*) informational privacy; *iv*) data sharing and autonomy; *v*) consent and the uncertain value of data; *vi*) ownership and data access; *vii*) social isolation; *viii*) decontextualisation of health and well-being; *ix*) care quality and user well-being; and *x*) risks of non-professional care. Berman and Cerf [60] comment on rights to privacy in the IoT and accountability for decisions made by autonomous systems. Hence, the ethical properties of algorithms need to be investigated. Mittelstadt et al. [61] categorise the ethical aspects of algorithms into *xi*) quality of evidence, divided into inconclusive, inscrutable, and misguided evidence; as well as *xii*) unfair outcomes; *xiii*) transformation effects; and *xiv*) traceability.

B. Regulatory Issues

Our presented framework consists of a variety of technical devices that are supposed to support the patient. As these devices and algorithms potentially could put the patient at risk, we need to consider the regulatory issues regarding medical devices. According to Boulos et al. [62], both the EU and the U.S. regulations define what a *medical device* is and whether certifications (e.g., according to the European Medical Device Directive MDD 93/42/ECC) would be required. Apps that archive and retrieve data from, e.g., patient records, support decisions informed by medical databases representing known facts, or perform straight-forward simple calculations are usually not considered as medical devices. In contrast, apps for diagnosis, dosage calculation, interpretation and interpolation of assessed data outside a clinician's supervision need to be certified as medical devices. Then, developers of such apps need to undertake a controlled test and risk assessment.

Considering the apps developed in our project for patients with axial SpA, the current implementation offers assessment of subjective and objective data, calculates indices according to the published definitions, transfer data to a health cloud, and presents trends for the patient to make a decision. According to the above considerations, these apps are not considered as medical devices. Note also that the patient recognises potential problems also without the apps; immediate danger for life due to the conditions cannot be expected.

We also note that the device for clinical measurements presented in Section IV-D would need certification as a medical devices since it is used in a clinical setting.

C. Ethics and Risk Analysis

We evaluate the fourteen ethics characteristics *i* to *xiv* for the currently implemented prototype. TABLE IV shows elements of this informal ethics assessment. According to this assessment, there are no critical issues; however, the implementation and deployment of the self-management system needs to be made carefully, and the issues mentioned in the assessment table need to be addressed.

We perform an informal risk analysis for our system that is similar to the work by Leister et al. [63]. From a system model, e.g., the one presented in Fig. 3, we can extract the functionality, communication channels, and list possible risks, sources of failure, weaknesses, exploits, and attacks.

TABLE IV: Informal ethics assessment for characteristics i to xiv.

#	Ethics aspect	Properties and comments
i	Personal privacy	Make sure that only authorised persons can access device.
ii	Obtrusiveness, stigma, autonomy	Apps and sensors can be used in private space; smartphones and sensors are not obtrusive.
iii	Informational privacy	Make sure that information security measures are implemented.
iv	Data sharing	Data sharing only when client permits; controlled by regulations.
v	Consent	Patient can give consent.
vi	Data ownership, data access	Ownership by patient regulated by law in Europe; data access by regulations.
vii	Social isolation	Reduces number of clinical consultations; it is unlikely that this will lead to social isolation.
viii	Decontextualisation	Patient initiates questionnaires and assessments; patient can delete unwanted values; however, mishaps can occur.
ix	Care quality	Patients are under clinical supervision, but number of clinical consultations is reduced.
x	Non-professional care	Patients are under clinical supervision.
xi	Quality of evidence	Machine learning is not yet applied; system gives recommendations to educated patients; when patients enter wrong data or perform exercises in wrong manner, results and recommendations will be affected; hardware-dependence possible.
xii	Unfair outcomes	n/a
xiii	Transformation effects	Health-care system will change; number of planned consultations will be reduced in the long run.
xiv	Traceability	Yes, for current implementation; weaker when machine learning is applied.

For these, we consider severity and probability, as well as countermeasures. One could formalise this even more as demonstrated by Leister et al. [64], but we consider this not necessary for an overview.

Risks can occur due to various reasons, including the following: *a*) attacks on information security and privacy, affecting both devices and communication channels; *b*) issues with technical safety when operating the devices; *c*) issues with correctness of decisions and results; *d*) malfunction of device or algorithms; *e*) operation errors caused by the user; *f*) reduced availability; and *g*) organisational issues and policies.

D. Analysis of the MOSKUS Apps

Since the self-management system is complex, the results shown by the device might be wrong. For example, wrong input values might result in displaying an incorrect status and wrong recommendations. This, in turn, might result in non-optimal treatment. For recommendations for axial SpA, this will not cause life-threatening situations. Further, the patients using the MOSKUS apps are supposed to be educated to manage their own disease without the app and will seek for medical assistance also when technology should fail.

Reasons of the apps showing misguiding results might be the following:

1) When using the questionnaires, the patient might enter wrong data. This may happen either voluntarily (cheating) or involuntarily by accident. In either case, when wrong values are entered, the resulting indices – such as ASDAS, BASDAI or BASFI – will be biased, leading eventually to wrong recommendations.
To avoid incorrect data, one can make a sanity check of the entered values and ask the patient when unusual data are entered. Such mechanisms have not been implemented in the prototypes, but this would be necessary for a final product. Further, educating the patient in the purpose and use of the apps would have a positive impact.

2) When performing the exercises, the patients might perform these in a wrong manner; e.g., they might move parts of the body that are supposed to be fixed or they might have misunderstood the exercise entirely. If this happens, the measurements from the exercises and the resulting indices (e.g., BASMI) will be biased.
To avoid this, educating the patients might be essential. Further, mechanical aids and fixtures, as well as sanity check of the exercises using extra sensors or recording more values might be necessary. Moreover, some patterns in movements during the exercises might indicate that exercises are not performed correctly. Such patterns can be detected in a preprocessing step to avoid submitting erroneous data. In the current prototypes, such measures have not been implemented.

3) Malfunction of the device or its sensors may happen. E.g., when using the smart phone as a measurement device for the exercises: the accelerometer could experience noise, the gyroscope could drift or be inaccurate, or the magnetometer might be influenced by magnetic fields. Indeed, in the tests carried out in this work, we observed that older smart phones may not be as accurate as required.
To avoid this, calibration steps and sanity checks of the sensors and devices need to be performed regularly and, if possible, without the intervention of the patient. If this is not possible, the apps should give out warnings in case of strange results during such calibration steps.
The consequences from malfunction of sensor or device will be no or unusable measurement values. This might result in incorrect recommendations given by the app.

There are further risks that can arise. Breaches of information security and privacy might compromise the privacy of the patient. Further, exploits might alter or destroy data and have an impact on the integrity of the data. As a consequence, this might lead to wrong recommendations or system failure (e.g., denial of service).

Suggested counter-measures are similar to those described by Leister et al. [63], i.e., to secure the communication channels by using security functions such as authentication, authorisation, access control, and encryption. Access to devices and sensors, as well as to the health cloud and the patient records by unauthorised persons needs to be restricted using the suitable measures. Note that security and privacy policies developed by the stakeholders are necessary, as for every other IT system.

Data loss can happen due to various reasons, e.g., unavailable services, destroyed hardware, power outage, theft, protocol errors, or routines not performed according to policies. The usual countermeasures for these situations apply, for example backing up important data.

Although smart phones and sensors can be considered as personal devices, it can happen that other persons, e.g., friends and family can get access to the MOSKUS app. For example, friends and family might be interested in the technology and try the app. To avoid that the data set is contaminated with data from other persons, the app needs to offer functionality to remove such data sets both from the smart phone, health cloud, and patient records. Additionally, a sanity check of the data can detect such unwanted data and suggest them for removal.

To summarise, the above informal analysis covers the some major risks that could lead to unusable data, privacy breach, data loss, and data inconsistency. However, there are several measures that can help avoid many of these risks becoming a threat to the patient. These measures include *a*) analysis of collected data and evaluation whether these data fit into the patient's usual pattern; *b*) analysis of the quality of exercises using extra sensors and data analysis; *c*) implementation of warnings when data are inconsistent or do not fit into the patient's pattern; *d*) possibility to remove data sets that are inconsistent, inaccurate or are form other sources than the patient; *e*) implementation of suitable security protocols; and *f*) possibility to back up data. Such measures and functionality need to be added before starting productification of the MOSKUS prototype.

VII. Discussion

The proposed concept for self-management is based on a feedback loop which involves the patient. Axial SpA does not require immediate attention when the condition worsens, but an appointment with a clinic needs to be scheduled. Also, not adhering to the self-management regime does not have other side-effects beyond not adhering to the treatment, and these patients need to keep the conventional frequency of clinical follow-ups. Note that other chronic diseases might require immediate attention in some situations or not adhering to the self-management regime might worsen the patient's condition. Thus, an evaluation is needed for other conditions than axial SpA to see if our self-management architecture can be applied.

Data assessed in self-management are usually not complete or might be of a different nature in terms of the clinical indices. For example, the values extracted from blood samples might not be available, only selected values from the BASMI might be available, or the patient assesses alternative measurements that are not part of the established indices. To support recommendations in these cases, it is necessary to predict an individual's axial SpA disease condition based on a combination of physiological, behavioural and subjective (self-reported) features. To achieve this, Schiboni et al. [65] have proposed a fuzzy rule-based evidential reasoning (FURBER) approach for multiple assessment fusion. But this approach requires enough real patient data as training data to be considered for real treatment.

The medical indices for axial SpA and the data retrieved from the FURBER method are only suited to give an indication of the disease conditions at one moment. For predicting the probable development of the patient's health condition and whether actions need to be taken requires temporal reasoning. Modelling the disease development as a stochastic process to optimise the treatment recommendations could be done by a Markov Decision Process (MDP) [66]. Yet a large sample size could make this approach less viable [67]. Alternatively, the *patient profiling* method described by Lutz et al. [68], could be feasible.

The new assessment methodology for rotation exercises using sensor technology will also impact clinical use as it will save time and provide better results. Today, health personnel use goniometer or compass-based measurements that have acceptable accuracy. The trials in a clinic have shown this new methodology simplifies clinical measurements, greatly improves accuracy. The time saved and higher-quality data quickly make up for the cost of the sensors. Specifically, the much higher accuracy and easier handling of the sensor technology compared to the traditionally used methods is attractive to health professionals. Furthermore, the sensor will enable patients to perform the measurements themselves without the involvement of health personnel.

The use of sensor technology that comes in today's smart phones for the assessment of the rotation exercises is viable. However, one has to keep in mind that the hardware is not standardised, and some models or brands could behave unexpectedly. Technical limitations might also have an impact on the implementation of the exercises, as our study has shown.

Our prototype does not implement all of the exercises that are recommended, and more studies and research will be necessary to find a suitable implementation of these. For some of these, e.g., the stair climb test or the sit-to-stand test, a solution seems obvious, unlike for others (e.g., the extended Schober's test). Thus, replacement-exercises that are viable for implementation using sensors that are affordable need to be developed and clinically approved.

VIII. Conclusion

We presented an architecture for self-management of axial SpA patients that is based on self-assessment by these patients. We have performed a proof of concept by implementing vital parts of a self-management system including clinical measurements, patient-reported outcome measurements, feedback module, patient diary, and decision making software. For this, we used both specifically designed sensors that are suitable for clinical settings as well as sensors that are on-board of smartphones.

Further user evaluations will be necessary before a system based on our architecture can be brought into clinical practice. In addition, communication modules to the EHR system of the clinics need to be implemented. Further, the development of suitable measurements for exercises beyond rotation exercises need to be developed in a way that allows patients to perform these at home. For some of the objective assessments, it must be considered to use replacement-exercises. This, however, will require evidence in clinical trials that these exercises are similar to the ones they replace.

Finally, since patient-reported data might not be of the best quality (e.g., they have not undergone quality assurance or might be incomplete) estimation methods both for the current disease status and for temporal prediction need to be developed. While we could show the viability of the methods, further implementation work needs to be done.

IX. Acknowledgments

Parts of the work presented here has been carried out in the **M**obile muscul**O** **SK**eletal **U**ser **S**elf-management (MOSKUS) project funded by the Research Council of Norway in the VERDIKT programme, grant number 227251.

References

[1] W. Leister et al., "Towards evidence-based self-management for spondyloarthritis patients," in Proc. Fifth Intl. Conf. on Global Health Challenges, pp. 43–48, Venice, Italy, Oct. 2016.

[2] L. C. Li, A. F. Townsend, and E. M. Badley, "Self-management interventions in the digital age: New approaches to support people with rheumatologic conditions." Best Pract Res Clin Rheumatol, vol. 26, no. 3, pp. 321–33, 2012.

[3] K. R. Lorig, P. D. Mazonson, and H. R. Holman, "Evidence suggesting that health education for self-management in patients with chronic arthritis has sustained health benefits while reducing health care costs," Arthritis & Rheumatism, vol. 36, no. 4, pp. 439–446, 1993.

[4] T. J. Brady, J. Kruger, C. G. Helmick, L. F. Callahan, and M. L. Boutaugh, "Intervention programs for arthritis and other rheumatic diseases," Health Education & Behaviour, vol. 30, no. 1, pp. 44–63, February 2003.

[5] M. Battersby et al., "Twelve Evidence-Based Principles for Implementing Self-Management Support in Primary Care," Joint Commission Journal on Quality and Patient Safety, pp. 561–570, Dec. 2010.

[6] Y. Miedany, "e-rheumatology: are we ready?" Clinical Rheumatology, vol. 34, no. 5, pp. 831–837, 2015.

[7] D. Palmer and Y. E. Miedany, "Proms: a novel approach to arthritis self-management," British Journal of Nursing, vol. 21, no. 10, pp. 601–607, 2012.

[8] Ø. Dale, T. Schulz, A. Christie, and H. Dagfinrud, "Collection of patient reported outcome measures using short messaging service," in eTELEMED 2013, The Fifth International Conference on eHealth, Telemedicine, and Social Medicine, pp. 214–219. IARIA, 2013.

[9] J. Braun and J. Sieper, "Ankylosing spondulitis," Lancet, vol. 369, pp. 1379–1390, 2007.

[10] M. Dougados and D. Baeten, "Spondyloarthritis," The Lancet, vol. 377, no. 9783, pp. 2127–2137, 2011.

[11] J. Braun et al., "2010 update of the ASAS/EULAR recommendations for the management of ankylosing spondylitis," Annals of the Rheumatic Diseases, vol. 70, no. 6, pp. 896–904, 2011.

[12] D. van der Heijde et al., "2016 update of the ASAS-EULAR management recommendations for axial spondyloarthritis," Annals of the Rheumatic Diseases, vol. 76, no. 6, pp. 978–991, 2017.

[13] P. West, R. Giordano, M. van Kleek, and N. Shadbolt, "The Quantified Patient in the Doctor's Office," in Proceedings of the 2016 CHI Conference on Human Factors in Computing Systems - CHI '16, pp. 3066–3078. New York, USA: ACM Press, 2016.

[14] S. Newman, L. Steed, and K. Mulligan, "Self-management interventions for chronic illness," The Lancet, vol. 364, no. 9444, pp. 1523–1537, Oct. 2004.

[15] S. Johnston, C. Liddy, S. M. Ives, and E. Soto, "Literature review on chronic disease self-management," Élisabeth Bruyère Research Institute, report, 15 April 2008, submitted to The Champlain Local Health Integration Network.

[16] U. E. Bauer, P. A. Briss, R. A. Goodman, and B. A. Bowman, "Prevention of chronic disease in the 21st century: elimination of the leading preventable causes of premature death and disability in the usa," The Lancet, vol. 384, no. 9937, pp. 45–52, 2014.

[17] K. R. Lorig et al., "Chronic disease self-management program: 2-year health status and health care utilization outcomes," Medical care, vol. 39, no. 11, pp. 1217–1223, 2001.

[18] J. Barlow, C. Wright, J. Sheasby, A. Turner, and J. Hainsworth, "Self-management approaches for people with chronic conditions: a review," Patient Education and Counseling, vol. 48, no. 2, pp. 177–187, 2002.

[19] M. Aitken, J. Lyle et al., "Patient adoption of mHealth – use, evidence and remaining barriers to mainstream acceptance," IMS Institute for Healthcare Informatics, report, Sep. 2015.

[20] C. You et al., "Soberdiary: A phone-based support system for assisting recovery from alcohol dependence," in Proceedings of the 33rd Annual ACM Conference on Human Factors in Computing Systems, CHI 2015, Seoul, Republic of Korea, April 18-23, 2015, B. Begole, J. Kim, K. Inkpen, and W. Woo, Eds., pp. 3839–3848. ACM, 2015.

[21] H. Brendryen, F. Drozd, and P. Kraft, "A digital smoking cessation program delivered through Internet and cell phone without nicotine replacement (happy ending): Randomized controlled trial," J Med Internet Res, vol. 10, no. 5, p. e51, Nov 2008.

[22] F. Drozd, L. Mork, B. Nielsen, S. Raeder, and C. A. Bjørkli, "Better days – a randomized controlled trial of an Internet-based positive psychology intervention," The Journal of Positive Psychology, vol. 9, no. 5, pp. 377–388, 2014.

[23] E. Årsand, M. Muzny, M. Bradway, J. Muzik, and G. Hartvigsen, "Performance of the First Combined Smartwatch and Smartphone Diabetes Diary Application Study," Journal of Diabetes Science and Technology, vol. 9, no. 3, pp. 556–563, May 2015.

[24] A. van Wissen, C. Vinkers, and A. van Halteren, "Developing a virtual coach for chronic patients: A user study on the impact of similarity, familiarity and realism," in Persuasive Technology, pp. 263–275. Springer, 2016.

[25] A. R. P. Azevedo, H. M. L. de Sousa, J. A. F. Monteiro, and A. R. N. P. Lima, "Future perspectives of smartphone applications for rheumatic diseases self-management," Rheumatology international, vol. 35, no. 3, pp. 419–431, 2015.

[26] H. S. Park, H. Cho, and H. S. Kim, "Development of a multi-agent m-health application based on various protocols for chronic disease self-management," Journal of medical systems, vol. 40, no. 1, pp. 1–14, 2016.

[27] K. A. Stroetmann, T. Jones, A. Dobrev, and V. N. Stroetmann, "eHealth is worth it – the economic benefits of implemented eHealth solutions at ten European sites," Office for Official Publications of the European Communities, report, 2006.

[28] M. Jagoda, "SpA Helper," web pages. [Online]. Available: https://itunes.apple.com/cv/app/spa-helper/id722932133 [Accessed: 2. May, 2016].

[29] X. Su, H. Tong, and P. Ji, "Activity recognition with smartphone sensors," Tsinghua Science and Technology, vol. 19, no. 3, pp. 235–249, June 2014.

[30] S. Alley, S. Schoeppe, D. Guertler, C. Jennings, M. J. Duncan, and C. Vandelanotte, "Interest and preferences for using advanced physical activity tracking devices: results of a national cross-sectional survey," BMJ Open, vol. 6, no. 7, 2016.

[31] P. Pires, L. Mendes, J. Mendes, R. Rodrigues, and A. Pereira, "Integrated e-healthcare system for elderly support," Cognitive Computation, vol. 8, no. 2, pp. 368–384, 2016.

[32] R. W. Picard, "Emotion research by the people, for the people," Emotion Review, 2010.

[33] W. Leister, I. Tjøstheim, G. Joryd, T. Schulz, A. Larssen, and M. de Brisis, "Assessing visitor engagement in science centres and museums," Journal on Advances in Life Sciences, vol. 8, no. 1&2, pp. 49–63, 2016.

[34] J. S. Smolen et al., "Treating spondyloarthritis, including ankylosing spondylitis and psoriatic arthritis, to target: recommendations of an international task force." Ann Rheum Dis, vol. 73, no. 1, pp. 6–16, 2014.

[35] S. A. Raymond and R. N. Ross, "Electronic subject diaries in clinical trials," Applied Clinical Trials, pp. 48–58, March 2000.

[36] J. Zochling, "Measures of symptoms and disease status in ankylosing spondylitis: Ankylosing Spondylitis Disease Activity Score (ASDAS), Ankylosing Spondylitis Quality of Life Scale (ASQoL), Bath Ankylosing Spondylitis Disease Activity Index (BASDAI), Bath Ankylosing Spondylitis Functional Index (BASFI), Bath Ankylosing Spondylitis Global Score (BAS-G), Bath Ankylosing Spondylitis Metrology Index (BASMI), Dougados Functional Index (DFI), and Health Assessment Questionnaire for the Spondylarthropathies (HAQ-S)," Arthritis Care Res.(Hoboken.), vol. 63 Suppl 11, pp. S47–S58, Nov 2011.

[37] K. Irons and C. Joffries, "The Bath Indices – outcome measures for use with ankylosing spondylitis patients," The National Ankylosing Spondylitis Society (NASS), Surrey, UK, 2009.

[38] J. Sieper et al., "The assessment of SpondyloArthritis international society (ASAS) handbook: a guide to assess spondyloarthritis," Annals of the Rheumatic Diseases, vol. 68, no. Suppl 2, pp. ii1–ii44, 2009.

[39] N. Østerås, H. Dagfinrud, A. T. Tveter, D. M. Soldal, C. Fongen, and C.-H. Franke, "About spondyloarthritis, disease specific questionnaires and assessments, and generic field tests for health-related physical fitness," Diakonhjemmet Sykehus, NKRR-notat 2-2013, 2013.

[40] T. Coppetti et al., "Accuracy of smartphone apps for heart rate measurement," European Journal of Preventive Cardiology, vol. 0, no. 0, pp. 1–7, 2017, doi: 10.1177/2047487317702044.

[41] A. Anjum and M. U. Ilyas, "Activity recognition using smartphone sensors," in 2013 IEEE 10th Consumer Communications and Networking Conference (CCNC), pp. 914–919, Jan 2013.

[42] A. Avci, S. Bosch, M. Marin-Perianu, R. Marin-Perianu, and P. Havinga, "Activity recognition using inertial sensing for healthcare, wellbeing and sports applications: A survey," in 23th International Conference on Architecture of Computing Systems 2010, pp. 1–10, Feb 2010.

[43] K. Zhan, F. Ramos, and S. Faux, "Activity recognition from a wearable camera," in 12th Int'l. Conf. on Control Automation Robotics Vision (ICARCV), pp. 365–370, Dec 2012.

[44] W. Zhao et al., "A human-centered activity tracking system: Toward a healthier workplace," IEEE Transactions on Human-Machine Systems, vol. 47, no. 3, pp. 343–355, June 2017.

[45] M. Tousignant, L. Poulin, S. Marchand, A. Viau, and C. Place, "The Modified Schober Test for range of motion assessment of lumbar flexion in patients with low back pain: A study of criterion validity, intra- and inter-rater reliability and minimum metrically detectable change," Disability and Rehabilitation, vol. 27, no. 10, pp. 553–559, 2005.

[46] American Thoracic Society, "ATS statement: guidelines for the six-minute walk test," Am J Respir Crit Care Med, vol. 166, no. 1, pp. 111–117, 1. July 2002.

[47] G. A. V. Borg, "Psychophysical bases of perceived exertion," Medicine and Science in Sports and Exercise, vol. 14, no. 5, pp. 377–381, 1082.

[48] E. J. Nightingale, F. Pourkazemi, and C. E. Hiller, "Systematic review of timed stair tests," JRRD, vol. 51, no. 3, pp. 335–350, 2014.

[49] C. Fongen, A. T. Tveter, F. R. Andersen, N. Østerås, and V. Berre, "Laboratory validity and reliability of a new sensor developed for measuring rotation around the vertical axis – a subproject in the Mobile mosculO SKeletal User Self-management," Diakonhjemmet Sykehus, NKRR-notat 1-2016, 2016.

[50] C. Fongen, N. Østerås, F. R. Andersen, V. Berre, A. T. Tveter, and K. B. Hagen, "Innovation in health care – laboratory validity and reliability of a new sensor developed to measure cervical rotation," Annals of the Rheumatic Diseases, vol. 75, no. suppl 2, p. 1288, 2016.

[51] J. Qi, L. Chen, W. Leister, and S. Yang, "Towards knowledge driven decision support for personalized home-based self-management of chronic disease," in Proc. of the 2015 Smart World Congress, pp. 1724–1729, 2015.

[52] P. Machado et al., "Ankylosing spondylitis disease activity score (asdas): defining cut-off values for disease activity states and improvement scores," Annals of the rheumatic diseases, vol. 70, no. 1, pp. 47–53, 2011.

[53] J. Braun et al., "International ASAS consensus statement for the use of anti-tumour necrosis factor agents in patients with ankylosing spondylitis." Annals of the rheumatic diseases, vol. 62, no. 9, pp. 817–824, 2003.

[54] G. Wariaghli et al., "Normative values for the Bath Ankylosing Spondylitis Functional Index in the general population compared with ankylosing spondylitis patients in Morocco," BMC Musculoskelet Disord, vol. 13, no. 1, p. 40, 2012.

[55] M. J. Kviatkovsky et al., "The minimum clinically important improvement and patient-acceptable symptom state in the BASDAI and BASFI for patients with ankylosing spondylitis," The Journal of Rheumatology, vol. 43, no. 9, pp. 1680–1686, 2016.

[56] J. Brooke, "SUS: A quick and dirty usability scale," in Usability evaluation in industry, P. W. Jordan, B. Weerdmeester, A. Thomas, and I. L. McClelland, Eds. London: Taylor and Francis, 1996.

[57] K. Finstad, "Response interpolation and scale sensitivity: Evidence against 5-point scales," J. Usability Studies, vol. 5, no. 3, pp. 104–110, May 2010.

[58] K. Ashton, "That 'Internet of Things' Thing," RFID Journal, 22 July 2009. [Online]. Available: http://www.rfidjournal.com/article/view/4986 [Accessed: 1. June 2015].

[59] B. D. Mittelstadt, "Ethics of the Health-Related Internet of Things: Mapping the debate," SSRN, 2017. [Online]. Available: https://ssrn.com/abstract=2932221 [Accessed: 15 June, 2017].

[60] F. Berman and V. G. Cerf, "Social and ethical behavior in the Internet of Things," Commun. ACM, vol. 60, no. 2, pp. 6–7, Jan. 2017.

[61] B. D. Mittelstadt, P. Allo, M. Taddeo, S. Wachter, and L. Floridi, "The ethics of algorithms: Mapping the debate," Big Data & Society, vol. 3, no. 2, p. 2053951716679679, 2016.

[62] M. N. K. Boulos, A. C. Brewer, C. Karimkhani, D. B. Buller, and R. P. Dellavalle, "Mobile medical and health apps: state of the art, concerns, regulatory control and certification," Online Journal of Public Health Informatics, vol. 5, no. 3, p. 229, 2014.

[63] W. Leister, H. Abie, A.-K. Groven, T. Fretland, and I. Balasingham, "Threat assessment of wireless patient monitoring systems," in Proc. ICTTA'08, Damascus, Syria, 2008.

[64] W. Leister, M. Hamdi, H. Abie, S. Poslad, and A. Torjusen, "An evaluation framework for adaptive security for the IoT in eHealth," International Journal on Advances in Security, vol. 7, no. 3&4, pp. 93–109, 2014.
[65] G. Schiboni, W. Leister, and L. Chen, "Using fuzzy evidential reasoning for multiple assessment fusion in spondylarthropathic patient self-management," in Emerging Trends and Advanced Technologies for Science and Information, ser. Studies in Computational Intelligence. Springer, 2016.
[66] C. C. Bennett and K. Hauser, "Artificial intelligence framework for simulating clinical decision-making: A markov decision process approach," Artificial Intelligence in Medicine, vol. 57, no. 1, pp. 9–19, 2013.
[67] A. J. Schaefer, M. D. Bailey, S. M. Shechter, and M. S. Roberts, "Modeling medical treatment using markov decision processes," in Operations Research and Health Care: A Handbook of Methods and Applications, M. L. Brandeau, F. Sainfort, and W. P. Pierskalla, Eds., pp. 593–612. Boston, MA: Springer, 2004.
[68] W. Lutz, Z. Martinovich, and K. I. Howard, "Patient profiling: An application of random coefficient regression models to depicting the response of a patient to outpatient psychotherapy." Journal of Consulting and Clinical Psychology, vol. 67, no. 4, p. 571, 1999.

List of Acronyms

ANT+	*an interoperable wireless transfer capability*
API	application program interface
AS	Ankylosing Spondylitis
ASAS	Assessment of SpondyloArthritis international Society
ASDAS	Ankylosing Spondylitis Disease Activity
aSpA	axial Spondyloarthritis
BASFI	Bath AS Disease Activity Index
BASFI	Bath AS Functional Index
BASG	Bath AS Patient Global Score
BASMI	Bath AS Metrology Index
BCI	brain-computer interface
CI	confidence interval
CST	chair stands test
CRP	C-reactive protein
EHR	electronic health record
ESR	erythrocyte sedimentation rate
EULAR	European League Against Rheumatism
EU	European Union
FURBER	fuzzy rule-based evidential reasoning
ICC	interclass correlation coefficient
ICT	information and communication technology
IoT	Internet of Things
ISM	industrial, scientific, and medical
LED	light emitting diode
LoA	limits of agreement
MDP	Markov decision process
mHealth	mobile health
MCII	minimum clinically important improvement
MOSKUS	mobile musculo-skeletal user self-management
PASS	patient-acceptable symptom state
PC	personal computer
RPE	rating of perceived exertion
SDC	smallest detectable change
SpA	Spondyloarthritis
ST	stair climb test
STD	standard deviation
SUS	system usability scale
U.S.	United States
6MWT	six minutes walking test

Personalised Health Monitoring by a Multiagent System

Leo van Moergestel, Brian van der Bijl,
Erik Puik, Daniël Telgen
Department of Computer science
HU Utrecht University of Applied Sciences
Utrecht, the Netherlands
Email: leo.vanmoergestel@hu.nl

John-Jules Meyer
Intelligent systems group
Utrecht University
Utrecht, the Netherlands
Alan Turing Institute Almere, The Netherlands
Email: J.J.C.Meyer@uu.nl

Abstract—**By using agent technology, a versatile and modular monitoring system can be built. This paper describes how such a system can be implemented. The roles of agents in this multiagent system are described as well as their interactions. The system can be trained to detect several combinations of conditions and react accordingly. This training can be done with specific patient situations in mind, resulting in a personalised monitoring system. Because of the distributed nature of the system, the concept can be used in many situations, especially when combinations of different sensor inputs are used. Another advantage of the approach presented in this paper is the fact that every monitoring system can be adapted to specific situations by varying the number and types of sensors and the messaging capability. As a case-study, a health monitoring system will be presented.**

Keywords–Multiagent-based health monitoring; Multiagent architecture; learning agent.

I. Introduction

The work in this paper is based on a paper presented at the Intelli 2016 conference [1] and other previous work. Monitoring systems are widely used in many situations. Simple systems collect information that can be inspected by humans or other systems. More advanced systems have the capability to react on the data monitored. Smoke detecting systems with an alarm are examples of these systems. Often a situation arises where more than one monitored condition should be taken into account before an action should be performed. Industrial production systems are examples of complicated situations where many sensors are used to control the process [2]. Another example of a complicated situation is the health condition of the human body [3]. Here, alarm conditions may also depend on individual factors, necessitating the monitoring system to be trained for the specific individual person.

This paper describes a modular agent-based system [4] that can be trained by a medical expert and can monitor the status of a person and react adequately on the conditions encountered. This system has been built using agent technology, resulting in a robust and versatile multiagent-based monitoring system. The concepts presented here can be used in other situations as well [5].

The rest of this paper is organised as follows: Section II will describe the concepts of our approach, the reason for choosing agent technology as well as the architecture of the multiagent system. The agent types introduced in the architecture desription as well as other technical aspects will be explained in more detail in subsecquent sections, starting with Section III where the design of the sensor agent is described. The decision agent is the subject of Section IV. Communication and the communication agents are explained in Section V and Section VI. The section is followed by Section VII where the training system will be explained. This training aspect is an important aspect of the system and is treated in detail. The proof of concept and results are presented in Section VIII. Related work will be discussed in Section IX and a conclusion will end the paper.

II. Agent-based monitoring

The first part of this section will show the requirements and explain the use of agent technology, while the second part will focus on the multiagent architecture.

A. Requirements and technology

A monitoring system should be built to be capable to handle several input values in combination. Depending on the combined values of the inputs a specific action should be executed. The system should be trained to build a knowledge base and utilise known information to decide on a strategy to react to the current situation. This resulted in the following list of requirements:

- the system should monitor different inputs simultaneously;
- it should be easy to add extra monitoring inputs;
- the system should be trained in an effective manner;
- the system should have a set of possibilities to react on certain conditions;
- different types of reaction should be possible depending on the input values.

As a case study, a system in the medical domain has been adopted, but the concepts presented here can easily be used in other domains as well.

For the realisation, agent technology has been used [4]. The reasons for choosing agent technology are:

- Error resistance. By using separate agents, the failure of an agent responsible for sensor input will not bring down the whole monitoring system. There is now a possibility to fall back on a different solution based on the availablity of sensor inputs.
- Clear separation of responsibilities and goals. In our design, the sensors will be tied to separate agents that have a clearly defined goal. This is also true for the other agents involved, as will be discussed in the next subsection.
- Modularity. A multiagent system (MAS) is modular by nature and can be easily expanded with new features and possibilities.

B. *MAS design*

The agents involved have roles and responsibilities. When the different responsibilities are taken into account this will result in the architecture of a multiagent system as depicted in Figure 1.

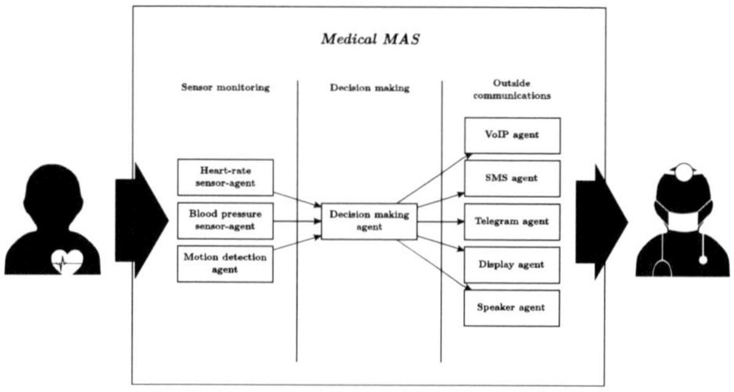

Figure 1. Medical MAS architecture

Three different roles are incorporated in the design resulting in three types of agents.

1) monitoring agents: Monitoring agents are responsible for delivering data to the decision agent. The data is coming from sensors. The agents themselves have a rather simple design. It could be possible to tell the agent at what intervals the data should be presented as well as the format expected by the decision agent.

2) decision making agent: A central role in the system is played by the decision agent. This agent decides what action should be performed under what conditions given by the monitoring agents. To do so, it has to be trained to build a knowledge base on how to react on certain conditions. This training has to be supervised by an expert, in our case a medical expert. A data acquisition system has been developed to help the expert to efficiently add data to the knowledge base. That system will be explained in the next section.

3) communication agent: The system has a set of communication agents that are responsible to communicate with the outside world. These agents are used by the decision agent to send emails, messages for several communication systems, like SMS, and also putting information on a display or generating an audible alarm.

Each MAS could contain any number of agents from the first and third categories (collectively known as utility agents), as well as one central agent having the capability of mapping observations from sensor agents to actions performed by communication agents. This design allows for agents to be added and removed dynamically while keeping core functionality intact. An example medical MAS is shown in Figure 1, including a number of potential utility agents.

C. *Internal communication*

The categories of agents described above will operate together in a single MAS. Though it would be possible to have agents running outside this MAS and still be able to communicate with the agent within, this aspect falls outside of the scope of this research.

The gold standard for agent development is the JADE developed by Telecom Italia. It provides a set of tools and an extendable framework for creating agents and MASs using the FIPA standard. Within JADE, agents exchange information using the ACL protocol [6] developed by FIPA. The prototype medical MAS has been developed in JADE and as such uses the ACL protocol for communicating information. The ACL protocol allows for a number of languages to encapsulate data, including XML; as the XML format provides a means to represent data in a semantic way that is readable to both humans and computers it appears to be a good choice for inter-agent communication. A typical exchange of messages is shown in Figure 2. Two sensor-agents send a stream of measurements to the decision agent, which periodically calculates its assessment of the situation. When the assessment reaches a certain defined threshold, it starts to send messages to the communication agents. The communication agents can respond with a confirmation or a message indicating either failure to relay the communication or failure to understand the message. If the decision-agent receives a message indicating failure, it tries to alert an operator using the designated fallback.

III. Sensor Agents — Biological factors and monitoring

Sensor-monitoring agents are the simplest and most diverse agents in the medical MAS. These agents exist primarily to support modularity: as sensors do not necessarily produce compatible signals to communicate their results, a small, dedicated agent could be programmed to read the sensor-data and communicate it to the decision-making agent in a standard format. This way, the decision-making agent does not need to know the way measurements are performed. A sensor could easily be exchanged by a different kind of sensor, together with its associated agent, without necessitating mayor changes to the decision-making agent. Similarly, new sensors could be added to facilitate patients with diseases requiering certain types of sensors. Though this would require new data to be added to the decision-making agent, the process of acquiring the measurements could easily be added by inserting a new sensor-monitoring agent to the MAS. This part of the system will be discussed in Section III.

In order for any system to be able to assess a patients health, it needs to know about a number of biological factors. For heart failure and related problems — the focus

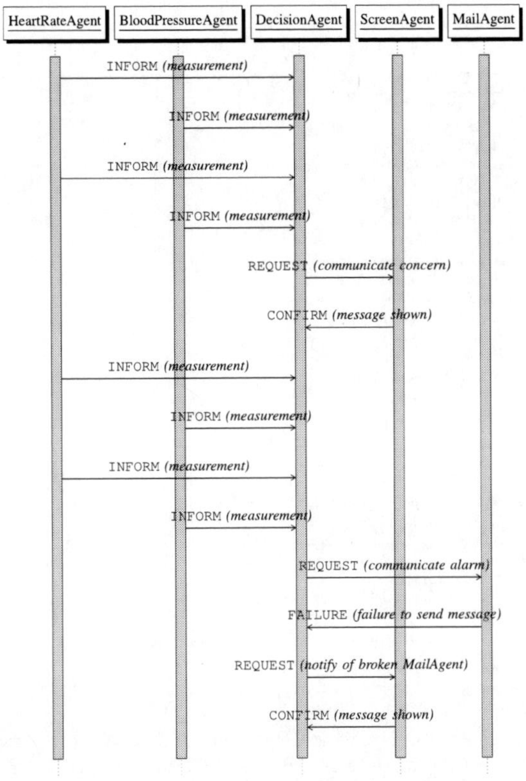

Figure 2. Message exchange within the medical MAS.

of this research, the patients heart rate and blood pressure seem to be obvious choices for factors to be monitored. For other diagnoses another set of variables might contain more meaningful information. However, from an artificial intelligence (AI) point of view, the provision of a decisive list of factors is not in the AI domain but in the medical domain. As such, this resulted in a design of a prototype that is as factor-agnostic as possible. Each factor to be considered is added during initialisation as a feature to a n-feature algorithm, and an appropriate sensor is added to the system to collect the necessary data. By scaling each feature to a value in the range $(-1, 1)$, as described in Section VII-H, the specifics of each feature are abstracted away from the reasoning process: the product does not need to know what a given value represents, only how it affects the output of its prediction-function.

Thus, for the system to work with a given set of variables, three things are needed to add the desired behaviour to the existing product:

1) a sensor capable of measuring the new feature,
2) a mapping between sensor-output and a value in the range $(-1, 1)$ and
3) a dataset for the prediction algorithm featuring the new factor.

To facilitate the development of sensor-agents, an abstract class has been written to minimise the required amount of boilerplate code. All sensor-agents have a similar structure: a single, repeating behaviour and a standardised method of communicating measurements to the decision-making agent. The `SensorAgent` class, as shown in Figure 3, provides an abstraction of these similarities and only needs a `setup()` method and a `performMeasurement()` method to be implemented to create a concrete class. The `setup()` method should set any relevant variables (of which the variables required by the `SensorAgent` class can be set by calling the `setup(String, Double, Double, String)` methods of the superclass). The `performMeasurement()` method should contain any code needed to perform a measurement, and return its results as a `Double`. The default behaviour of the `SensorAgent` is to execute the measurement-method every 2 seconds and to send the measurement to the decision-making agent using an XML message as shown in Listing 1. The number of 2 seconds can be changed according to the situation and type of measurements to be made. In the example, a subclass called `HeartRateAgent` sends a message containing both the raw values of the measurement (a real number within the range supported by the sensor) and the same measurement scaled to the range of $(-1, 1)$.

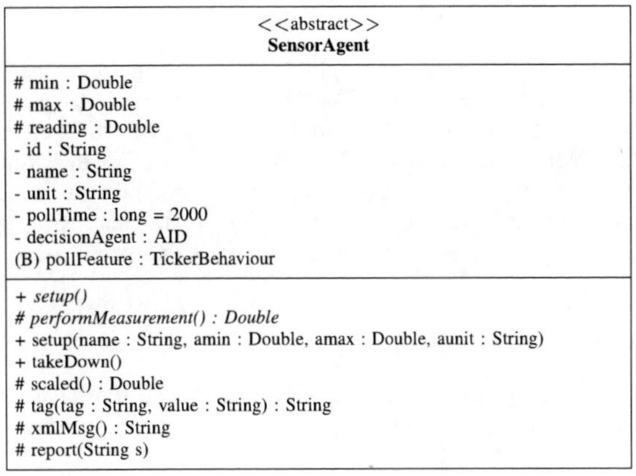

Figure 3. The SensorAgent abstract class.

Listing 1. A typical message sent from `HeartRateAgent`

```
<measurement>
  <feature>HeartRate</feature>
  <raw>68.950161</raw>
  <scaled>-0.080665</scaled>
</measurement>
```

The `SensorAgent` itself contains a number of private and protected attributes to store its specifics such as the name it uses for communication, the specifics of the associated sensor such as minimum / maximum / current values, and the Agent ID of the decision agent. A `SensorAgent` contains a single behaviour, represent here as (B).

As described above, the `SensorAgent` contains two abstract functions needed to implement a subclass. In addition, the class contains a few helper-methods to limit code duplication. The method `scaled()` uses the minimum and maximum values to apply feature scaling (described in more detail in Section VII-H), `tag(String, String)` and `xmlMsg()` provide a more readable way to generate the XML and `report(String)` is a wrapper around `System.out.println(String)` prepending the output with the agent name to make it easier to distinguish messages when various agents are reporting at the same time.

A. *Example: HeartRateAgent*

As an example, Listing 2 provides a template for how a `HeartRateAgent` would look as a subclass of `SensorAgent`. The hardware-specific code can be filled in when a sensor has been selected to produce a functional sensor-agent.

Listing 2. A sample SensorAgent subclass

```
public class HeartRateAgent extends SensorAgent
{ public void setup()
  { super.setup("HeartRate", 0.0, 150.0, "bpm"); }

  public Double performMeasurement()
  { try
    { // read sensor
      // calculate heart rate from reading
      return reading; } // feature scaling happens
    catch (Exception e) //    when message is sent.
    { // handle exception
      return reading; } } } // return old reading
```

IV. The Decision Agent — Agent Reasoning

The decision agent is the central part of our MAS, and is responsible for mapping measurements to assessments of the patient's situation and to request external communication if the situation becomes dire. The agent is trained to perform this task using logistic regression on datapoints provided by a medical expert. The result of this training procedure is a variable $\boldsymbol{\theta}$, which is unique for every patient. Its structure is dependent on the number of factors monitored and the amount of feature-mapping applied, and its contents vary slightly to account for differences between patients, even when the structure is identical. The $\boldsymbol{\theta}$ variable and the level of feature-mapping are stored within the agent, and will periodically be used to transform the vector of measurements $\boldsymbol{x}$ into a prediction value representing the certainty the system has of the patients well-being.

In order to make the agent as generalised as possible it will be given a set of general behaviours dependent on variables such as $\boldsymbol{\theta}$, ensuring that most of its behaviour can be changed by sending updated parameters instead of changing the code. The agent will need more information than just the $\boldsymbol{\theta}$ variable: it will need a list of "plans" telling it how to react to certain situations. Furthermore, the agent needs to know about the order of the variables within x, the level of feature-mapping, and which agents to contact.

A UML overview of the decision-agent's structure is shown in Figure 4. As shown, the decision agent contains a great amount of variables and operations. The agent contains

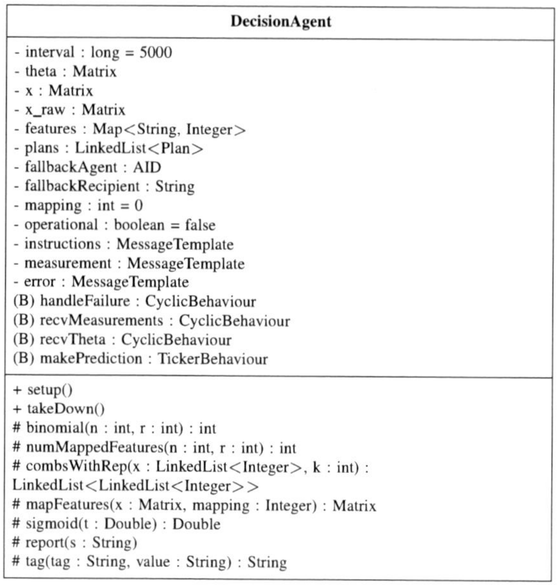

Figure 4. The decision-agent class.

matrices (or more precisely, mathematical vectors) for storing $\boldsymbol{\theta}$ and $\boldsymbol{x}$ (the latter both scaled and as raw values). A `Map` is used to associate the names of features with their position in `x`. In addition, it contains a variable `interval` controlling how often a prediction is made, an integer telling the feature mapping function how many polynomials it should generate, a set of `MessageTemplates` allowing it to distinguish various kinds of messages, and a `boolean` indicating whether the agent is operational (*i.e.*, has received a valid set of instructions). The list of `Plans` and the fallback communication method are described below in Section IV-D.

A. *Methods*

In addition to the `setup()` and `takedown()` methods required by JADE as pseudo-constructors / destructors, the agent contains a number of utility-functions serving to make the code (mostly contained in its behaviours) more readable and to improve maintainability. `binomial(..)`, `combsWithRep(..)`, `numMappedFeatures(..)` and `mapFeatures(..)` are used to perform and verify the feature mapping on the agent.

B. *Behaviours*

As the decision-agent is the central part of the medical MAS, it contains a large set of behaviours: three cyclic behaviours, which are constantly active, and a ticker behaviour operating on an interval depending on the `interval` variable:

- `handleFailure` listens for messages indicating failure in any of the communication agents. In such an event, it will use a designated fallback-agent to alert an operator that the system might be unable to communicate.

Plan
- below : Double - message : String - recipient : String - agent : AID - limit : Integer - available : boolean = true - timer : Timer
+ Plan(b : Double, m : String, r : String, a : AID, t : Integer) + toString() : String + execute(h : Double) - msg_x() : String - xmlMsg() : String

Figure 5. The plan inner class

- `recvMeasurements` listens for messages from the sensor-agents and saves them in `x` and `x_raw` for use in predictions.
- `recvTheta` listens for messages containing instruction sets. This aspect is explored in Section IV-E.
- `makePrediction` is responsible for periodically multiplying `theta` and `x` to make a prediction regarding the patient's health. This process is described in Section IV-C.

C. Assessing the situation

Every `interval` milliseconds, the agent uses its then-current knowledge of the features, represented in `x`, to construct a feature-mapped column-vector `x_mapped`. The inner product of `x_mapped` and the row-vector `theta` is passed through the `sigmoid(double)`-method yielding a `double` between zero and one, representing the probability that the patient is still healthy. As this number decreases, the probability of something being wrong increases. After a prediction has been calculated, the value is compared to the thresholds defined for each plan; if the calculated result is below the threshold for a given plan, the agent will attempt to execute it by messaging a communication agent.

D. Executing plans

Plans are represented by a special `Plan` class, shown in Figure 5. Each time a prediction is made, the agent attempts to invoke the `execute(Double)` method for each plan, passing the predicted probability. Each plan contains a threshold `below`, which is compared to the prediction when `execute(Double)` is called. Execute will send its message to its specified recipient via its specified agent, provided two conditions are met:

1) The prediction passed as an argument to `execute(Double)` is lower than or equal to the threshold for the plan and
2) The `boolean available` is set to `true`

The value of `available` is initialised as `true`, but is set to false when the plan is first executed. At the same time, a timer is started for `limit` seconds, after which `available` is reset to `true`. This prevents the MAS from flooding its recipients with messages as a new prediction is calculated, by default, every five seconds; though it may be meaningful to provide an occasional update, a realistic poll frequency for the agent to make predictions is likely always higher than a realistic notification frequency. By using a plan specific interval all frequencies can be chosen separately.

E. Receiving instructions

All of the necessary information can be delivered to the agent within a single ACL message; Listing 3 shows a sample XML fragment containing instructions for a decision-agent using two features, including a second-degree feature mapping and two plans.

Listing 3. An initialisation message as sent to the decision-agent.

```
<instructions>
  <features>
    <feature id="SystolicBloodPressure">
      <label>Systolic Blood Pressure</label>
      <min>0</min>
      <max>200</max>
      <unit>mm Hg</unit>
    </feature>
    <feature id="HeartRate">
      <label>Heart Rate</label>
      <min>0</min>
      <max>200</max>
      <unit>bpm</unit>
    </feature>
  </features>
  <mapping>2</mapping>
  <theta>
    <value>2.402548</value>
    <value>2.769392</value>
    <value>3.467782</value>
    <value>-7.500590</value>
    <value>-2.189613</value>
    <value>-11.995721</value>
    <value>-2.301167</value>
    <value>2.064028</value>
    <value>-2.568114</value>
    <value>-2.736256</value>
  </theta>
  <plans>
    <plan>
      <below>0.6</below>
      <message>Watch out!</message>
      <via>ScreenAgent</via>
      <to></to>
      <limit>30</limit>
    </plan>
    <plan>
      <below>0.4</below>
      <message>Panic!</message>
      <via>MailAgent</via>
      <to>brian.vanderbijl@hu.nl</to>
      <limit>3600</limit>
    </plan>
  </plans>
  <fallback>
    <via>ScreenAgent</via>
    <to></to>
  </fallback>
</instructions>
```

The initialisation consists of five parts:

1) A `features` node containing information about each feature. The order the features are presented in determines the position of measurements within x, and must be identical to the order of features during the learning process.

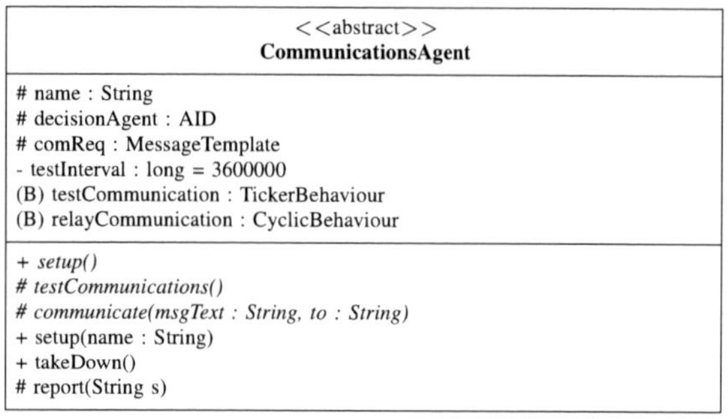

Figure 6. The CommunicationsAgent abstract class.

2) A `mapping` node containing a single integer value determining the amount of feature-mapping.
3) A `theta` node containing a series of decimal values representing $\boldsymbol{\theta}$. As with the features, the order is important here.
4) A `plans` node containing a set of plans used to respond to predictions. Each plan includes a threshold, below which the plan will be executed, a message to be sent, the name of the agent responsible for relaying the message, an optional recipient (whether this is needed depends on the agent: a mail or SMS agent would require a recipient, whereas a screen agent would not), and a limit in seconds determining how often a plan can be executed in order to avoid flooding messages.
5) A `fallback` node containing an agent and a recipient to alert when a communications agent is not functioning properly and cannot be trusted to relay important messages.

When the decision-agent receives a set of instructions, it will confirm whether the length of $\boldsymbol{\theta}$ matches the length of $\boldsymbol{x}$ after feature scaling, throwing an error if the two are incompatible. As each level of feature-mapping adds a number of features equal to $\binom{n+d-1}{d}$, the following equality must hold:

$$\text{size}_\theta = \sum_{i=1}^{d} \binom{\text{size}_x + i - 1}{i} .$$

V. Communicating Results to the Outside World

The decision-agent as described in Section IV relies on other agents to communicate the results to a medical expert and/or the patients themselves. This choice is deliberate, as it allows new methods of communication to be added "on the fly", without changing the decision-agent's behaviour. Each communication agent added to the system represents a new option to communicate the patient's health and relay concern. As with the sensor-agents, an abstract class has been provided to facilitate the development of additional agents. This class, `CommunicationsAgent`, shown as UML in Figure 6.

Each communication agent has a `String` variable to hold its name and an `AID` representing the decision agent. A variable `testInterval` controls how often the agent performs a self-diagnostic. Two behaviours are present: one to continually listen for requests for communication, and another to perform the self-tests on an interval dictated by `testInterval`. The class provides the methods for setup, agent destruction and reporting to `stdout`.

Three abstract methods need to be implemented to create a `CommunicationsAgent` subclass:

1) `setup()` should set any relevant variables, at the very least including the agent's name.
2) `testCommunications()` should include the code needed to run a self-test, if applicable, and throw an exception if it fails to complete the test. This exception is caught by the `testCommunication` behaviour after which a `FAILURE` message is sent to the decision-agent indicating the communication agent has become unreliable.
3) `communicate()` should include all code needed to send a message, such as setting up the necessary objects for IO in Java (provided this needs to be done each time a message is sent; if the method of communication features a persistent object that can be trusted to remain operable, it can be setup in the `setup()` method) and actually sending the message. It will send a `CONFIRM`-message back to the decision agent if the sending process did not encounter any errors; in case of failure it can send either a `NOT_UNDERSTOOD` message to indicate the XML received was illegible, or a `FAILURE` indicating some sort of IO error encountered in trying to relay the message to its recipient.

VI. Communications Agents — Requests for Outside Communciations

Requests for communication from the decision agent are packaged in a small snippet of XML, as shown in Listing 4. The `message`-node contains the body of the email, including the most recent value for each feature.

Listing 4. A typical message sent to `MailAgent`

```
<request>
  <to>leo.vanmoergestel@hu.nl</to>
  <message>
    Patient health in serious condition!
    - HeartRate = 54.93483905942176
    - SystolicBloodPressure = 86.20808412990199
  </message>
</request>
```

Just as there are various ways to acquire data to facilitate the decision-making agent, there are also many methods to communicate its results. These include telephony, instant messaging, patient-information logs and on-device IO like displays, alarms, etc. The preferred methods of communication may be subject to change over time as the patient's situation changes, as doctors come and go and as new forms of communication are developed, become widely adopted and are eventually deprecated. Therefore, MAS communication to the outside world should be modular. Just like with sensor-reading agents described in Section III, methods of communications could be implemented by small, trivial single-purpose agents.

By abstracting the way information is delivered, the decision-making agent can communicate its results in a predefined manner indicating the conclusions to be sent and the perceived level of panic. Communications agents can pick up these messages and assume responsibility of relaying the information to the appropriate recipients.

Using the abstract class approach, any method of communication can be added to link the system up to existing medical care, and an existing system can easily be extended to include new ways of communicating. Depending on the availability of usable Java libraries this may be done in relatively small, simple agents.

VII. Data aquisition for agent training

In order to interpret the measurements acquired from the sensors and predict whether the current patient situation constitutes a cause of alarm, the decision agent needs a way to classify potentially high-dimensional data. Each biological factor considered in the model represents an additional dimension for data points. As this information is not guaranteed to be available for various combinations of biological features, it makes sense to explore a way for medical personnel to easily enter such data into the system. Not only does this guarantee the required data can be generated, if not available, it also allows for far greater personalisation, providing the agent with a data-set tailored to its patient. Manual entry, or at least confirmation, also allows an expert intimate knowledge of the agents decision-making process, potentially increasing trust by removing the "black box" aspect of machine learning.

Teaching the system to recognise alarming measurements and differentiate between various levels of threat requires large amounts of information provided by medical personnel, preferably tailored to the patient as thresholds might not be the same for every person. Entering this data can be challenging: as potentially multiple factors need to be taken into account together, it becomes progressively harder for humans to visualise and communicate relevant thresholds. A better way might be to input a set of data-points, together with appropriate assessments of the situation associated with each data-point. These data-points could be used, alone or in conjunction with more general datasets, to train a classification algorithm.

In order to train an agent to make accurate predictions, training data will need to be entered into the system by a medical expert. This should be as easy as possible: the focus should be to quickly train an agent without expending significant time accommodating the system. Unfortunately, entering possibly poly-dimensional data graphically is a difficult task. For one or two dimensional data, clicking points in a scatter plot, as pictured in Figure 7, can be a quick way to enter points; for three dimensional data this becomes harder: a scatter-plot is still possible for data-visualisation, but entry becomes impossible as a mouse or trackpad and a computer screen are both essentially two-dimensional. For even more simultaneous features, only a subset of the features can be plotted at the same time.

An alternative approach would be to require the expert to manually enter all features, as well as the results that the system should predict. Not only is this rather work-intensive, but also prone to omissions: as it is hard for the human mind to visualise all features simultaneously and large gaps are a significant risk.

A better solution would be for the system to dynamically suggest data-points based on the largest knowledge gaps. An expert would then be provided with the parameters for a new datapoint by the algorithm. For this datapoint an assessment of the situation can then be entered. The algorithm continuously updates its collection of datapoints, as well as the model derived from the combination of datapoints and expert assessments, and proceeds to suggest the largest empty areas in its knowledge-continuum as possible locations for new datapoints. This continues until the expert considers the fit of the model to be satisfactory, after which the model is accepted. The expert remains in control of the process of entering datapoints, and can at any time ignore a suggestion or opt to enter the parameters for a new datapoint themself.

This section considers an approach to accomplish this. Each problem will be examined in two dimensions first, as this makes it easier to visualise and demonstrate the applied methods. After the solution has been sufficiently exposed a generalisation can be made in n-dimensions.

To represent gaps in the knowledge-continuum, we create a triangulation of the known datapoints. Each datapoint is considered a vertex in an n-dimensional space, and by triangulating over this set of vertices we can detect sparsely populated areas by the emergence of larger triangles. In contrast, a large amount of datapoints in close proximity will yield a large number of smaller triangles.

Triangles and their higher-dimensional analogues (the tetrahedron in three dimensions, the 5-cell in four, etc.) are collectively referred to as n-simplex or just simplices (singular: simplex). As a triangle (2-simplex) is defined by three vertices of the form (x, y) and and a tetrahedron (3-simplex) is defined by four vertices of the form (x, y, z), an n-simplex is the most basic n-dimensional object defined by $n + 1$ vertices in n-dimensional space.

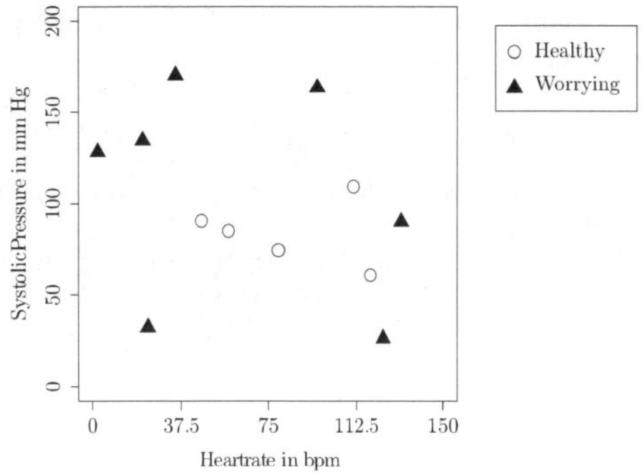

Figure 7. Scatter plot in two dimensions of a small random dataset.

A. Finding the most valuable points for data-querying

When entering data-points to train an agent, some points are more valuable than others. For example, potential locations completely surrounded by existing data-points all belonging to the same class are unlikely to add any new information to the system. Similarly, points in sparse areas are potentially more valuable, as are points closer to the centre of the point cloud. Figure 8 shows the same scatter plot as Figure 7, but adds a decision boundary and three possible locations for new data points marked by numbers. Location 1 does not appear to be a good addition, as it is very close to existing points and is therefore unlikely to add a great deal of information. Location 2 is not a good suggestion either, as it is very far from the decision boundary — it will likely have the same category as the points surrounding it, especially if a large amount of data has been entered. Location 3 is a better spot for a new data point: it is not a near duplicate of another point, and it lies close to the decision boundary. Depending on the category this point will be assigned to it may significantly change the decision boundary in either direction.

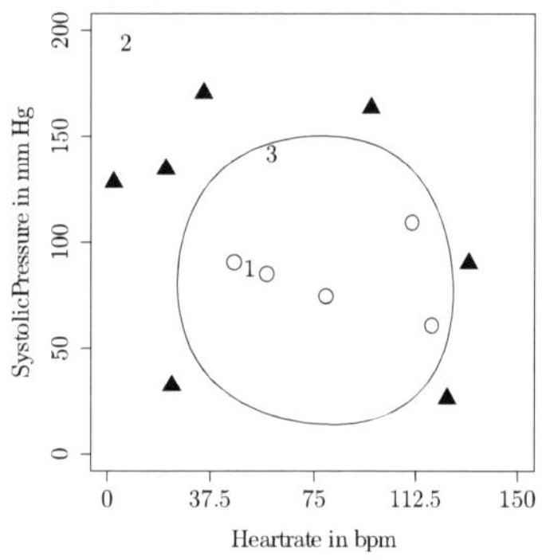

Figure 8. Three possible locations for a new data-point.

B. Data-point-distribution

To find sparsely populated areas to add new data-points, we first create a triangulation containing all data points. For each of these triangles, the circumcentre is calculated, and the collection is ordered based on the area of the triangles. These points can now be evaluated in order to find points close to the current decision-boundary.

C. Triangulating n-dimensional space in simplices

To triangulate a set of points we utilise the Delaunay Triangulation [7]. Most mathematical libraries include a function to quickly get the Delaunay Triangulation of a set of points in n dimensions. Triangulating the example data from Figure 7 yields the triangulation as shown in Figure 9.

D. Calculating the size of each n-simplex

To find the largest simplex we use the determinant of the matrix constructed by adding each vector representing a vertex as a single column, and adding a final row of ones [8]. For a triangle, the absolute value of the result is equal to two factorial times the triangle's area. For a tetrahedron, the absolute value

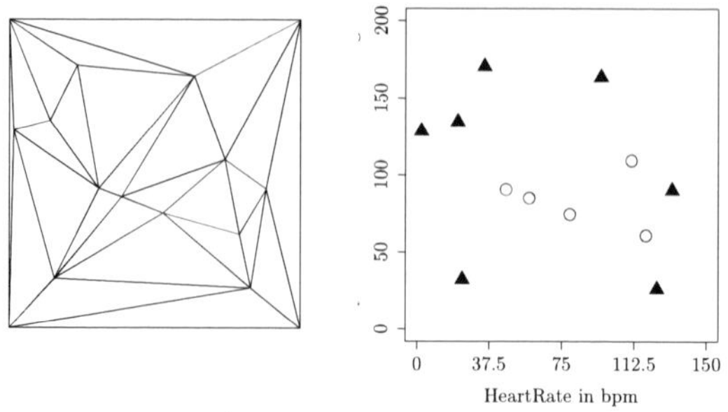

Figure 9. Triangulation and scatter plot in two dimensions

equals three factorial times the volume. For higher-dimensional shapes, this method continues to yield a scalar multiple of the n-hypervolume of the simplex. As the simplex size is only used for sorting, the scalar multiplication does not influence the ordering and can safely be ignored. As an example, the size of a triangle described by $a = (0, 0)$, $b = (0, 4)$ and $c = (3, 0)$ is given by

$$\text{abs}\left(\begin{vmatrix} 0 & 0 & 3 \\ 0 & 4 & 0 \\ 1 & 1 & 1 \end{vmatrix}\right) = 12 \quad (1)$$

which is twice the area of the triangle.

E. Calculating the circumcentre of each n-simplex

Once the largest data-gap has been found, we want to find its centre to suggest as a new data point. A simplex has multiple definitions of its centre; for this purpose the circumcentre, the point equidistant from all its vertices [9], seems a logical choice. Given a n-simplex defined by vertex $\boldsymbol{v}^{(1)}, \boldsymbol{v}^{(2)}, \ldots \boldsymbol{v}^{(n+1)}$ with a circumcentre c, we know that the distance between any vertex and c must, by definition, be equal. For any two vertices $\boldsymbol{v}^{(a)}$ and $\boldsymbol{v}^{(b)}$, this means:

$$\begin{gathered} \|\boldsymbol{v}^{(a)} - \boldsymbol{c}\| = \|\boldsymbol{v}^{(b)} - \boldsymbol{c}\| \\ \|\boldsymbol{v}^{(a)} - \boldsymbol{c}\|^2 = \|\boldsymbol{v}^{(b)} - \boldsymbol{c}\|^2 \\ (\boldsymbol{v}^{(a)} - \boldsymbol{c}) \cdot (\boldsymbol{v}^{(a)} - \boldsymbol{c}) = (\boldsymbol{v}^{(b)} - \boldsymbol{c}) \cdot (\boldsymbol{v}^{(b)} - \boldsymbol{c}) \end{gathered} \quad (2)$$

We translate each vector by $-\boldsymbol{v}^{(1)}$ so that $\boldsymbol{v}^{(1)}$ becomes the origin (denoted $\boldsymbol{o}$) and equate the distance to $\boldsymbol{c}$ of each remaining vector with the distance of $\boldsymbol{c}$ to $\boldsymbol{o}$, yielding the locus for each translated vertex $\boldsymbol{v}$ and the origin $\boldsymbol{o}$:

$$\begin{gathered} (\boldsymbol{o} - \boldsymbol{c}) \cdot (\boldsymbol{o} - \boldsymbol{c}) = (\boldsymbol{v} - \boldsymbol{c}) \cdot (\boldsymbol{v} - \boldsymbol{c}) \\ \boldsymbol{c}^2 = \boldsymbol{v}^2 - 2\boldsymbol{v} \cdot \boldsymbol{c} + \boldsymbol{c}^2 \\ 2\boldsymbol{v} \cdot \boldsymbol{c} = \boldsymbol{v}^2 \\ \boldsymbol{v} \cdot \boldsymbol{c} = 0.5\boldsymbol{v}^2 \\ \boldsymbol{v}_1\boldsymbol{c}_1 + \boldsymbol{v}_2\boldsymbol{c}_2 + \cdots + \boldsymbol{v}_n\boldsymbol{c}n = 0.5\|v\|^2 \end{gathered} \quad (3)$$

Doing this for every vertex $\boldsymbol{v}^{(2)}$ to $\boldsymbol{v}^{(n+1)}$ gives us n equations, allowing us to find the n-dimensional vector $\boldsymbol{c}$.

We can write these equations in matrix form and solve all equations simultaneously:

Writing

$$S = \begin{pmatrix} v_1^{(2)} - v_1^{(1)} & v_1^{(2)} - v_1^{(1)} & \dots & v_1^{(2)} - v_1^{(1)} \\ v_2^{(3)} - v_2^{(1)} & v_2^{(3)} - v_2^{(1)} & \dots & v_2^{(3)} - v_2^{(1)} \\ \vdots & \vdots & \ddots & \vdots \\ v_n^{(n+1)} - v_n^{(1)} & v_n^{(n+1)} - v_n^{(1)} & \dots & v_n^{(n+1)} - v_n^{(1)} \end{pmatrix}$$

$$\boldsymbol{c} = \begin{pmatrix} c_1 \\ c_2 \\ \vdots \\ c_n \end{pmatrix} \quad \boldsymbol{r} = 0.5 \begin{pmatrix} \|v^{(2)} - v^{(1)}\|^2 \\ \|v^{(3)} - v^{(1)}\|^2 \\ \vdots \\ \|v^{(n+1)} - v^{(1)}\|^2 \end{pmatrix}, \tag{4}$$

we have

$$S\boldsymbol{c} = \boldsymbol{r} \ . \tag{5}$$

Given this, we can multiply both sides by S^{-1} to get

$$\boldsymbol{c} = S^{-1}\boldsymbol{r} \ . \tag{6}$$

As c was translated by $-v^{(1)}$, all that remains is adding $v^{(1)}$ to find the triangle's circumcentre.

F. Avoiding suggesting out-of-bounds points

As shown in Figure 9, Delaunay triangulations are prone to yielding obtuse simplices, in particular around the edges. This can be a problem because an obtuse simplex has a circumcentre outside itself. On the edges, this will result in the algorithm suggesting points outside the sensor's bounds. As these points are meaningless and only serve to distract the user, we would like to avoid generating obtuse simplices.

We solve this problem by introducing a border of false data-points around the edge. These data-points are only used to determine the Delaunay triangulation, and are not present in the actual training-data being generated. The number of data-points is determined by a variable $\beta \in \mathbb{N}_1$: For $\beta = 1$, only the corners of the graph are added. For larger values of β, each axis is subdivided into β parts. As β becomes larger, out-of-bounds points become increasingly unlikely, and suggestions start to gravitate towards existing data-points.

As Figure 10 and Figure 11 show, too large a value for β makes the algorithm increasingly unlikely to suggest points around the edges. Though more central points are preferred, limiting data-points to a central cluster might not be the way to go. A solution for this could be to gradually decrease β over time.

G. Generating the borders

The set of points to be used as a border constitutes of the following:

- a point for each vertex of the n-cube describing the range of data-points

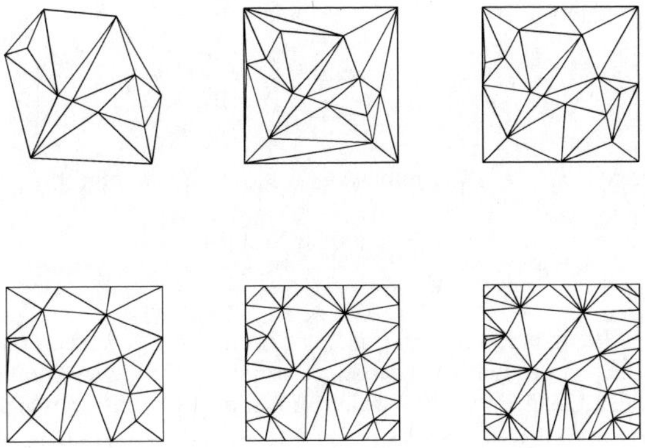

Figure 10. Triangulation for $\beta \in \{1, 2, 3, 8, 12\}$ alongside original triangulation.

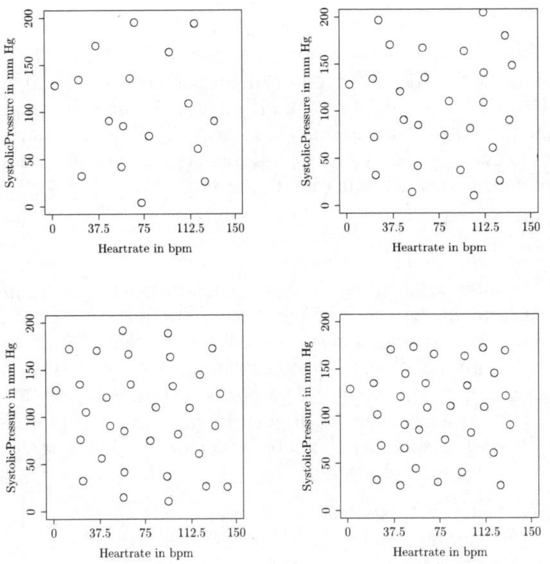

Figure 11. Scatter plot of the first twenty suggestions for $\beta \in \{1, 2, 4, 8\}$. Note that out-of-bounds points are not plotted.

- $(\beta - 1)$ points on each edge (1-face)
- $(\beta - 1)^2$ points on each face (2-face)
- $(\beta - 1)^3$ points on each cell (3-face)
- ...
- $(\beta - 1)^{n-1}$ points on each $(n-1)$-face

The number of points denoted by $\#P$ needed given a dimensionality n and a border-saturation β can therefore be calculated by

$$\#P(n, \beta) = \sum_{i=0}^{n-1} F(n, i)(\beta - 1)^i \quad (7)$$

where $F(n, i)$ is the number of i-faces on a n-cube [10]:

$$F(n, i) = 2^{n-i} \binom{n}{i} \quad (8)$$

The actual value of $P(n, \beta)$ can intuitively be seen as the Cartesian product of n instances of interval(β), also known as its Cartesian Power, of which only those points for which at least one of its members is equal to -1 or 1 are kept. In other words, for which the infinity norm $\|x\|_\infty$ equals 1.

$$P(n, \beta) = \{x \mid x \in \text{interval}(\beta)^n \wedge \|x\|_\infty = 1\} \quad (9)$$

$$\|x\|_\infty = \max_i |x_i| \quad (10)$$

H. Feature Scaling

The interval-function creates an interval between -1 and 1 in β steps. This is because all features are scaled to lie between -1 and 1, even though the actual measurements might range from 0 to some arbitrary maximum. This feature scaling is applied to make sure that all features are of the same importance when applying logit later on.

I. Avoiding symmetry

The algorithm presented above tends to favour generating a symmetrical data-set: As the range of values is a perfect n-cube, the first point suggested will be the centre, followed by a group of points equidistant from the first. This is undesirable, as symmetrical data points feature will introduce redundant features when multiplied during the fmap process. It will not help in generating a better hypothesis but will slow down the learning algorithm.

To prevent generating such a duplicate set of data, we will move each suggestion by a small random amount, controlled by a variable δ, that represents the maximal displacement for each point in each dimension. In order to ensure that this displacement will not place points outside the feature boundaries, this displacement will be opposite to the sign of the original location. This results in the data point being moved slightly towards the centre, which generally is the most interesting area to collect data on. We achieve this by replacing each vector element c_i by the weighted mean of $r \cdot 0$ and $(1-r)c_i$, where $r \sim U([0, \delta])$ is a random variable uniformly distributed on $[0, \delta]$.

VIII. Implementation

For the implementatation of the proof of concept, Java agent development framework (Jade) [11] has been used. The Jade runtime environment implements message-based communication between agents running on different platforms connected by a network. The reasons for choosing Jade are:

- the system presented is a multi-agent-based system. Jade provides the requirements for multiagent systems;
- the agent communication standard "Foundation for Intelligent Physical Agents" (FIPA) [12] is included in Jade;
- Jade is Java-based and it has a low learning curve for Java programmers; Java is a versatile and powerful programming language;
- Jade is developed and supported by an active user community.

The prototype has been developed and implemented on a standard Linux-based laptop. It should be possible to operate the system on any small device capable of running Java such as the Raspberry Pi nano [13]. Though the Jade-platform was selected for the prototype, this does not preclude development of a medical MAS in another framework or language. The concepts explored here can be implemented in any language, though support for a solid agent-development framework would be a serious asset. Nevertheless, if better performance is needed, the same principles could be implemented in a lower-level language, such as C, reducing much of the overhead at the cost of lower maintainability.

The prototype has been built and the working has been tested. In summary the following results have been achieved:

- The concept of a medical MAS consisting of three types of agents working together to monitor the patient and communicate the result.
- A method of collecting data from medical experts and utilising this knowledge to teach an agent to evaluate readings provided by sensors.
- The beginnings of a generalised framework upon which to build agents for inclusion in a medical MAS.

The next step will be implementing the system in the real world and testing the usability.

IX. Related work

Agent-based monitoring for computer networks has been proposed and implemented by Burgess. Burgess [14] [15] describes Cfengine that uses agent technology in monitoring computer systems and ICT network infrastructure. In Cfengine, agents will monitor the status and health of software parts of a complex network infrastructure. In [16], an agent-based monitoring system is proposed. A so-called product agent is responsible to monitor the working of a system in several different phases of its lifecycle. The actions performed by the agent are limited to prevent disasters or misuse. The aforementioned concept of a product agent that supports a product during its lifecycle from production to recycling is described in [17].

A lot of literature is available regarding health monitoring systems. Pantelopoulos and Bourbakis [18] give an overview of wearable sensor-based systems for health monitoring and prognosis. Their work focusses on the hardware implementation of the monitoring systems as well as communication technologies that might be used by such systems. The work of Milenkovic [19] is dedicated to wireless sensor networks in personal health monitoring. The system they describe collects data that is transferred to a central monitoring system whereas the system described in our paper aims for autonomous operation. Furthermore, monitoring systems that focus on special health related situations exist, such as the work of Marder et al. [20] where a system for monitoring patients with schizophrenia is described. An agent-based health monitoring as a concept for application of agent technology has been proposed by Jennings and Wooldridge in [21].

X. CONCLUSION

In this paper, a complex, expandable and agent-based monitoring system has been proposed and a proof of concept was built. The system turned out to work as expected. The design of the MAS has been described in detail as well as the communication between the different types of agents. Special attention has been given to the way the system builds its knowledge-base, resulting in an efficient system that focusses on the borders of operating space where transitions from one situation to another situation are possible. In the case of the medical monitoring system, this could result in a personal adapted monitoring system that can also be easily changed. Though the system is designed for use in a medical context, the concepts can be used in other domains as well.

REFERENCES

[1] L. v. Moergestel, B. v. d. Bijl, E. Puik, D. Telgen, and J.-J. Meyer, "A multiagent system for monitoring health," IARIA, Intelli The Fifth International Conference on Intelligent Systems and Applications, Barcelona, Spain, 2016, pp. 57–62.

[2] L. v. Moergestel, J.-J. Meyer, E. Puik, and D. Telgen, "A versatile agile agent-based infrastructure for hybrid production environments," IFAC Modeling in Manufacturing proceedings, Saint Petersburg, 2013, pp. 210–215.

[3] J. T. Parer and T. Ikeda, "A framework for standardized management of intrapartum fetal heart rate patterns," American Journal of Obstetrics and Gynecology, vol. 197, no. 1, 2007, pp. 26.e1 – 26.e6.

[4] M. Wooldridge, An Introduction to MultiAgent Systems, Second Edition. Sussex, UK: Wiley, 2009.

[5] L. v. Moergestel, J.-J. Meyer, E. Puik, and D. Telgen, "Embedded autonomous agents in products supporting repair and recycling," Proceedings of the International Symposium on Autonomous Distributed Systems (ISADS 2013) Mexico City, 2013, pp. 67–74.

[6] Foundation for Intelligent Physical Agents. FIPA ACL Message Structure Specification. [Online]. Available: http://www.fipa.org/specs/fipa00061/SC00061G.html

[7] Wolfram MathWorld. Delaunay Triangulation. [Online]. Available: http://mathworld.wolfram.com/DelaunayTriangulation.html

[8] P. Stein, "A note on the volume of a simplex," The American Mathematical Monthly, vol. 73, no. 3, 1966, pp. 299–301. [Online]. Available: http://www.jstor.org/stable/2315353

[9] Wolfram MathWorld. Circumcenter. [Online]. Available: http://mathworld.wolfram.com/Circumcenter.html

[10] R. J. McCann, "Cube face," 2010. [Online]. Available: http://www.math.toronto.edu/mccann/assignments/199S/cubeface.pdf

[11] Telecom Italia. JAVA Agent DEvelopment Framework. [Online]. Available: http://jade.tilab.com/

[12] Foundation for Intelligent Physical Agents. FIPA. [Online]. Available: http://www.fipa.org/

[13] E. Upton. Oracle Java on Raspberry Pi. [Online]. Available: https://www.raspberrypi.org/blog/oracle-java-on-raspberry-pi/

[14] M. Burgess, "Cfengine as a component of computer immune-systems,," Proceedings of the Norwegian Informatics Conference, 1998, pp. 283–298.

[15] M. Burgess, H. Hagerud, S. Straumnes, and T. Reitan, "Measuring system normality," ACM Transactions on Computer Systems (TOCS) Volume 20 Issue 2, 2002, pp. 125–160.

[16] L. v. Moergestel, J.-J. Meyer, E. Puik, and D. Telgen, "Monitoring agents in complex products enhancing a discovery robot with an agent for monitoring, maintenance and disaster prevention," ICAART 2013 proceedings, vol. 2, 2013, pp. 5–13.

[17] L. v. Moergestel, E. Puik, D. Telgen, and J.-J. Meyer, "The role of agents in the lifecycle of a product," CMD 2010 proceedings, 2010, pp. 28–32.

[18] A. Pantelopoulos and N. G. Bourbakis, "A survey on wearable sensor-based systems for health monitoring and prognosis," IEEE Transactions on Systems, Man, and Cybernetics, Part C (Applications and Reviews), vol. 40, no. 1, 2010, pp. 1–12.

[19] A. Milenković, C. Otto, and E. Jovanov, "Wireless sensor networks for personal health monitoring: Issues and an implementation," Computer communications, vol. 29, no. 13, 2006, pp. 2521–2533.

[20] S. R. Marder, S. M. Essock, A. L. Miller, R. W. Buchanan, D. E. Casey, J. M. Davis, J. M. Kane, J. A. Lieberman, N. R. Schooler, N. Covell et al., "Physical health monitoring of patients with schizophrenia," American Journal of Psychiatry, vol. 161, no. 8, 2004, pp. 1334–1349.

[21] N. R. Jennings and M. Wooldridge, "Applications of intelligent agents," in Agent technology. Springer, 1998, pp. 3–28.

Cardiac Telemerty for Stress Assessment

Anna Polevaia, Sergey Parin, Valeriia Demareva
Psychophysiological Department
UNN
Nizhni Novgorod, Russia
e-mail: a.dostoevskaya@gmail.com, parins@mail.ru,
kaleria.naz@gmail.com

Sofia Polevaia
Neurophysiology and Experimental Modeling Department
NNSMA
Nizhni Novgorod, Russia
e-mail: polevaia@ipfran.ru

Abstract—**The use of stress assessment is receiving increasing attention. However, assessment approaches and methods this far are limited to static contexts. Scholars have raised questions concerning the validity of laboratory measurement results for stress detection in everyday life. The paper presents the results of developing a new method of stress assessment in real-life scenarios. This method is based on the noninvasive continuous cardiac rhythm monitoring. It is shown that the results obtained in the laboratory stress-related contexts have poor correlation with the results of the similar real-life scenarios. In this connection, the principal requirements for stress monitoring solutions are listed, and the architecture of the cardiac telemetry complex for continuous stress measurement is described. Drawing on experimental data over a 5-year period, we identify a stress-specific context-invariant HRV pattern. The theoretical and application perspectives for cardiac telemetry are presented for discussion. The progress achieved in the research and the ways for further investigation are outlined in the conclusion.**

Keywords- stress; heart rate variability; autonomic nervous system; wireless technology.

I. Introduction

This paper is an extended presentation of [1], an experimental and theoretical justification for the role of endogenous opioid system in autonomic regulation of stress.

Stress reaction is a physiological response to factors perceived as challenging, demanding or threatening. Extensive scientific literature shows that stress can be associated with adverse health outcomes, e.g., an increased risk of severe chronic diseases, illnesses such as cardiovascular and immunological diseases - contributing to poor relationships and lost productivity at work, reduced quality and duration of life [2-4]. The overall burden of stress-generating lifestyles is heavy both at individual and at the societal level [5].

Measuring and monitoring stress is a key to identifying the main stressors and understanding the lifestyle-related choices and behaviors that make for healthy life and well-being. There is a dramatic growth of solutions for stress assessment over the past five years - from portable ambulatory diagnostics to home appliances, where the most advanced technologies investigate the potential of a modern smartphone and a wearable HR monitor [6] (see Figure 1). They allow to measure stress in a wide range of laboratory

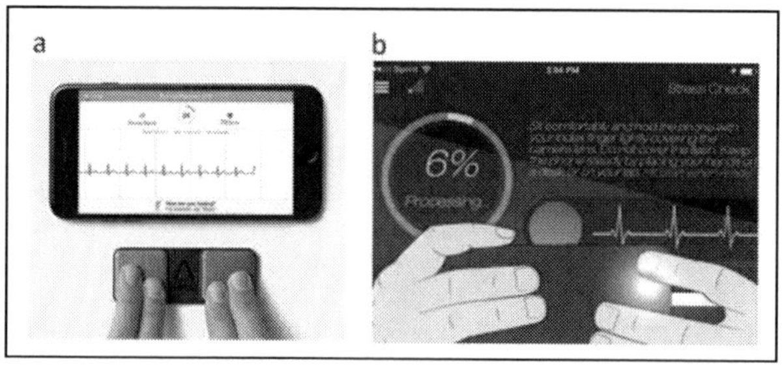

Figure 1. Modern approaches to cardiac activity monitoring: (a) a pocket-sized healthcare gadget (2017); (b) measuring stress through a built-in smartphone camera and flash (2015).

and ambulatory contexts, or static real-life scenarios like sleep. However, such approach greatly narrows the scope of application of the technology and questions its predictive validity, since the human regulation is a dynamic process.

This article introduces a cardiac telemetry solution for stress assessment in dynamic real-life scenarios and presents the results of its implementation in various experimental contexts. The developmental work is based on a combination of mathematical modeling and algorithm development with extensive empirical physiological and behavioral research. The datasets for developing the method include 7 432 field and laboratory assessments in 4 619 people. The technology described in this paper can be utilized widely in different types of healthcare studies, wellbeing and health promotion services, and consumer products such as chest belts, smart watches, as well as professional sports coaching and coaching amateur athletes.

The article is organized as follows: Section II presents a brief description of the stress phenomenon and introduces related works; it is followed by Section III, dedicated to the concepts and implementation details of our cardiac telemetry solution. Section IV addresses the evaluation results. Section V ends the paper with a look to future work.

II. Related Work

This section starts with a brief definition of the stress phenomenon and an overview on Heart rate variability (HRV) analysis approaches that are relevant in the context of stress detection. Finally, a short sub-section gives an overview on mathematical methods for stress modeling.

A. Definition of Stress

Stress is one of the most common extreme regimes for the living. According to the three-component theory, stress is a nonspecific protective systemic reduced response to damage or threat of damage [7-8]. That is, stress is not an adaptive, but a maladjustment process.

The stress-launching factor is a disagreement between the current state and the necessary state [9]. Then, in accordance with the theory of functional systems (FS) [10], we can assume that the occurrence of excessive mismatch in any of the modules of the acting FS leads to stress activation, which realizes the protective mechanism. The stress activation involves a combination of certain physiological (neurochemical, immune, vegetative, etc.) mechanisms [7-8, 11]. The same load can be optimal (or within the boundaries of the adaptation range) or stressing (beyond the adaptation range) for different people.

B. Stress Recognition using HRV

Autonomic nervous system (ANS) plays a key role in maintaining physiological functions of the body, including flexible and appropriate modification of the cardiac activity according to need. According to Baevsky et al., cardiac rhythm regulation can be represented as a double-circuit model with the central and autonomous circuits, with feedforward and feedback links (see Figure 2). In this case, the effect of the autonomous circuit is connected with respiratory arrhythmia, and the central circuit - with non-respiratory arrhythmia [12].

HRV analysis - determining the degree of variability in consecutive RR intervals or instant heart rate (HR) in a cardiogram - has become an important tool for assessing stress. A wide range of mathematical tools includes statistical, geometric, spectral, and nonlinear algorithms. Evolution of the mathematical apparatus for R-R intervals analysis is demonstrated by George E. Billman [13] (Figure 3). At the same time, the whole set of mathematical methods of HRV analysis is redundant [14]. However, it is important to note that each method has its own limitations in application. The most accurate method in terms of the time resolution is spectral analysis of RR-intervals. The selection of HRV periodic components allows us to determine the current state of the neuro-humoral regulation system. For the first time, different types of sequences of RR-intervals were described by Fleisch and Beckman in 1932 [15]. Rhythmic activity of pacemaker cells of the sinus node is interrelated with endocrine and humoral processes that change the threshold of spontaneous depolarization of pacemaker sinus node [16-18]. This leads to an increase or decrease in the interval between heart cycles and, consequently, a decrease or increase in heart rate. The factors that regulate HR will also affect HRV. An important feature of this process is that the activity of these factors varies periodically [19-20]. However, it should be noted that in addition to periodic modulations of various factors in HRV, there are also non-periodic components.

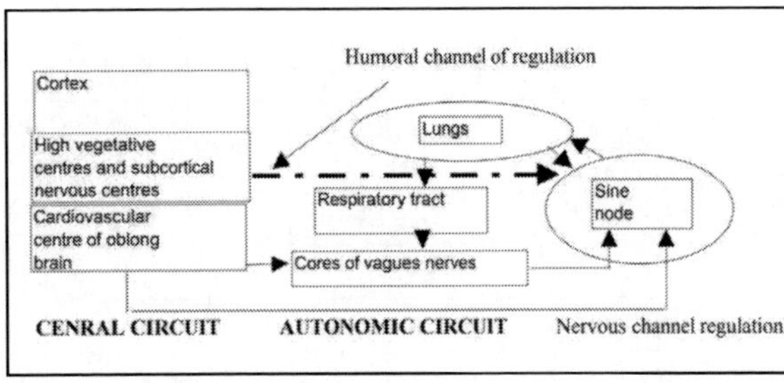

Figure 2. Scheme of double-circuit model for cardiac rhythm regulation.

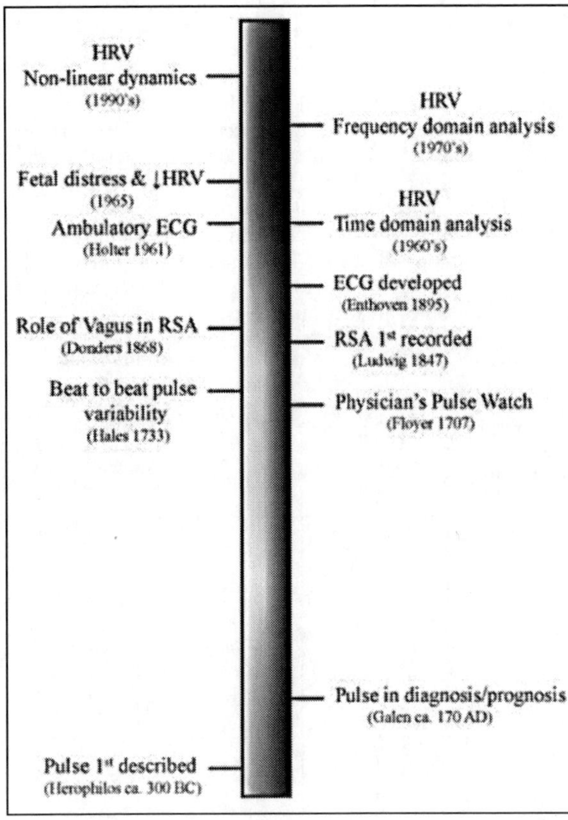

Figure 3. The temporal sequence of some of the most important events in HRV research history [13].

The traditional analysis of RR-intervals divides the HRV frequency spectrum into three ranges, (a) high-frequency (HF) oscillations (0.15-0.4 Hz); (b) low-frequency (LF) oscillations (0.04-0.15 Hz); (c) very low frequency (VLF) oscillations (0.015-0.04 Hz). However, in addition to the classical frequency ranges listed above, a number of authors indicate very high frequency (VHF) oscillations (0.6-2 Hz). Experimental data show the presence of VHF-components in the structure of the HRV spectrum in patients and in healthy people [21-22]. Table I shows the most relevant frequency ranges for spectral analysis of HRV.

TABLE I. THE FREQUENCY RANGES FOR SPECTRAL HRV ANALYSIS

Title of the Spectrum Component	Frequency Range (Hz)	Time Period (Sec)
HF	0,4 – 0,15	2,5 – 6,6
LF	0,15 – 0,04	6,6 – 25,0
VLF	0,04 – 0,015	25,0 – 66,0
ULF	Below 0,015	Over 66,0

The first studies of the impact of the information load on the functional state of a person using cardiovascular indicators are mentioned in the work of Winkler [23], who demonstrated that performing an arithmetic test leads to an increase in HR. Accumulated empirical evidence indicates that the LF spectrum is the most sensitive to the effect of cognitive loads. As a result, it was concluded that the suppression of LF (0.1Hz) HRV component reflects the effort of the subject required to perform a cognitive task, while the restoration of the spectral power during the relaxation period after completion of the cognitive task reflects the degree of previous efforts [24]. Some examples of empirical studies are listed in Table II.

TABLE II. EXAMPLES OF HRV SPECTRAL ANALYSIS UNDER COGNITIVE LOAD

Author	Experimental Problem	Selected HRV Indices	Trend
Egelund, 1982 [61]	Fatigue spectral HRV indices in drivers	LF (0,1 Hz)	Decrease
Kaplan, 1999 [62]	The change in the spectral indices in response to an erroneous action	HF (0,15-0,4 Hz)	Decrease
Mulder, 2000 [63]	The change in the spectral indices under cognitive load	LF (0,1 Hz)	Decrease
Brinkman, 2004 [64]	The change in the spectral indices during transition from solving simple equations to the complex ones	LF (0,1 Hz)	Decrease

The autonomic HRV regulation reflects the level of the adaptive resources of the organism [12; 25]. Studies have shown that emotions, cognitive processes, and physical activity are closely related to the dynamics of autonomic HRV regulation [26-27]. The research into stress provides an extensive HRV database for different groups of subjects in clinical and laboratory contexts: patients with depressive disorders, post-stroke patients with diabetes, etc. [28-30]. Vegetative correlates of fatigue, overexertion and various types of stress are actively studied [31-34]. Integration of the available data leads to the conclusion that the autonomic HRV regulation is sensitive to changes in emotional, cognitive and physical activity and is informative for the study of adaptation and maladjustment processes.

The data on the dynamics of HRV during stress are contradictory. Some studies demonstrate no changes in HRV [35] and the severity of respiratory arrhythmia in HR [36] in the presence of stress factors, while other authors show a high degree of variability in LF, HF, LF/HF spectral HRV parameters [37] (see Table III). Such contradictions could be explained by different software and hardware base of researchers.

TABLE III. SELECTED EMPIRICAL STUDIES OF HRV IN STRESS-RELATED CONTEXTS

Author	Context	LF Trend	HF Trend
T. Chandola et al., 2008 [34]	Working stress	Decrease	Decrease
J. Taelman et al., 2011 [26]	Intellectual load and/or physical load	Decrease	Decrease
N.I. Shlyk, 2011 [56]	An intensive training	Increase	Increase
H. Che-Hao et al., 2012 [57]	Inhalation anesthesia	Increase	N/D

For decades stress could be evaluated at rest only [38] with sitting or lying position of the subject being mandatory, and a great risk to damage the data by any external irritant stimuli [39]. All the mentioned restrictions greatly narrow the scope of contexts for research and make stress measurement in real-life scenarios impossible. Since the autonomic HRV regulation is dependent on the target function and varies greatly in accordance with the context [25, 40], static measurements in the laboratory do not always agree with the principle of ecological validity [15, 30, 41-42].

Currently, the problem of stress-specific HRV parameters is still unsolved. The definition of such parameters is complicated by the presence of individual optimum indices for a particular person, which do not always coincide with the average statistical results. This conditions the use of correlation of the HRV parameters along with the absolute values.

C. Neurohumoral HRV Regulation

More than 150 years ago, Claude Bernard denoted the existence of close connections between the brain and the heart [43]. Practical implementation of this concept is characterized by a number of important advantages. First, the methods of measuring the parameters of the functioning of the cardiovascular system (minute and stroke volume, pulse rate, arterial pressure) are well-known and generally available. Secondly, cardiovascular variability data can be used to evaluate the system of neurohumoral regulation of the heart and blood vessels, of which HR is the most simple and accessible for analysis.

Baro- and chemoreceptors control various parameters of the blood circulation in the vascular bed and heart, and as a result, the information about the ongoing endogenous changes enters the central nervous system (CNS). This ensures the lability of the adaptation of the heart-vascular system (HVS) to continuously changing environmental conditions. Thus, by controlling the processes of HVS regulation, it is possible to obtain information on the effectiveness of adaptive mechanisms in response to stress conditions.

As a theoretical approach, the neurovisual integrative model (NVM) is considered, according to which changes in HR are generally related to the state of the brain more, than to the state of the heart [44]. A number of nerve structures associated with heart rhythm are described within the framework of NVM. The data on these results are based on animal studies, studies of people with local cerebral lesions, physiological and pharmacological analysis, and work with methods of neuroimaging (PET, fMRT) [45-46].

HR is determined by the property of automatism, i.e., the ability of the cells of the conduction system of the heart to spontaneously activate and cause a contraction of the myocardium. Automatism is caused by the appearance of spontaneous depolarization of cells of the sinus node. The usual frequency is 60-80 pulses per minute. The fluctuations of HR are connected, on the one hand, with the intrinsic activity of the sinus node (intracardiac reflexes), and on the other hand with the influence of the higher centers of regulation [8]. HRV analysis, which is determining the degree of variability of consecutive R-R intervals or instant HR, has become an important tool for risk assessment.

HF oscillations of HR reflect the connection between the vagus nerve and the sinus node and the exerted neuromuscular influences. Therefore, the spectral power (density) in the HF range of the HRV spectrum is related to the activity of the parasympathetic link of the vegetative nervous system [11].

The LF range of the HRV spectrum when analyzing the records of the R-R intervals measured at rest (lying, sitting) is usually represented by a single peak at a frequency of 0.1 Hz. In fact, a wave peak with a frequency of 0.1 Hz in the spectrum of HRV means that the body has mechanisms for modulating HR with a period of 10 s. Oscillations with the same period are recorded in the rhythm of blood pressure. It can be assumed that the formation of a 0.1-Hz rhythm of RR-interval oscillations is the result of the participation of three mechanisms: baroreflex, central, and myogenic. For practical use, it is important that LF modulation of HR is associated with the activity of postganglionic sympathetic fibers, and their spectral power (density) reflects the activity of the sympathetic link of the VNS in the regulation of the heart rhythm [50].

As a rule, an increase in the power of LF oscillations is accompanied by a decrease in the power of HF oscillations, which may be the result of the existence of special mutual-reciprocal relations between them. Such interactions are also observed between the parasympathetic and sympathetic contours of the VNS, which determine the presence of these wave oscillations in HRV. This substantiates the usage of the LF/HF ratio power ranges of the HRV spectrum (also called the vegetative balance index, or VBI) for assessing the autonomic balance level [34, 39].

D. *Stress Research Issues*

The development of mathematical processing of ECG naturally led to the discovery of a large number of indicators (statistical, geometric, frequency), which, on the one hand, closely correlate with each other, making the entire set redundant, and on the other hand, are suitable for interpreting and evaluating the cardiac signal only in stationary conditions [14].

The key issues of autonomic regulation research are the following:

- What factors are responsible for autonomic responses in the context of daily-life activity?
- What psychophysiological processes underlie the HRV regulation in the context daily-life activity?
- Is it possible to predict HRV response in daily-life activity context from laboratory measurements?

For accessing physiological stress, a neurophysiologic index is a simultaneous increase of LF/HF values with the fall in the total power spectrum of HRV [11-12, 31, 33-35, 39-40].

The relationship between laboratory and field indices of vegetative regulation of the heart rhythm has been studied for more than 30 years.

Public speaking seems to be the most realistic natural activity context for stress study, along with an examination, an interview, etc. All of these real-life scenarios remain relatively manageable [42]. However, when public speech context was used as a component of the laboratory model of stress in the Trier Social Stress Test (TSST) [47], the later comparison of the results with 41 trials of natural public speaking HRV records showed no correlations [48]. Numerous studies show much more pronounced changes in HR when performing in public in the natural activity context as compared to laboratory experiments [49-51]. The review [41] concludes that there are weak correlations between HRV measurements produced in laboratory and field contexts, although some positive correlations are found.

Another approach to the problem of real-time stress diagnostics in natural dynamic contexts is the development of a new instrumental method. The method should allow recording of biophysical signals to provide personalized monitoring and remote diagnostics of stress without restrictions on the duration of recording. The California Institute of Technology and the Georgia Institute of Technology have made significant progress in this respect. They developed a clothing fabric with an incorporated network of contactless capacitive sensors to allow continuous HRV monitoring in real-life contexts [52].

In our work, we analyze HRV patterns from normal healthy people in various real-life contexts, e.g., stressful tasks, intensive physical activities, sudden maneuver while driving, etc. In order to do that, we use an unobtrusive HRV measuring device that can be easily worn long-term (24-hours). Such approach will allow us to identify the stress-specific HRV pattern for dynamic real-life scenarios.

III. CARDIAC TELEMETRY SOLUTION

The cardiac telemetry technology was designed for stress detection in dynamic real-life scenarios. The basic data for analysis are R-R intervals of the electrocardiogram (ECG) obtained in real time from a wireless sensor. The StressMonitor client application allows preliminary processing of the data and its representation in the form of a dynamic graph with indicated HRV spectral values. All the data is stored in a cloud. The solution is already in demand by medical institutions and scientific laboratories, professional coaches, and health-conscious people.

A. Solution Requirements

Since the operation of the human body is a dynamic process, due to high lability and the contextual dependence of HRV, special requirements are imposed on the technology for stress measurement in real-life scenarios. Among such requirements we list the following: safety and convenience for use in everyday life; mobility, i.e., the signal source can be remote from the signal receiving unit; continuous recording of signal; low power consumption; autonomous mode of HRV diagnostics; automatic processing of interrupts; sustainability to external interference; ability to accumulate data in off-line mode.

B. Solution Arcitecture

Cardiac telemetry is a monitoring solution that allows continuous HRV monitoring while the volunteer remains active without the restriction of being attached to a bedside cardiac monitor. A three-module wireless technology has been developed [53-55].

ZephyrTM HxMTM Smart (HxM, Zephyr Technology) sensor platform combines optimal size and energy consumption as well as quality of receiving and transmitting radio signals not sacrificing the comfort of the use. The platform is 65x32 mm in size and 17 grams in weight. Its design ensures reliable fixation of sensors on the human body. The platform combines a microprocessor, a radio signal reception and transmission unit, a low-power ECG sensor, an acceleration sensor, and a distance sensor. The device works for 150 hours without recharging. The signal transmission range is 10 m. The data transfer to a smartphone or a tablet PC is organized via Bluetooth SPP - 2,4 GHz. Packets of raw data are transmitted every second, where each dataset contains a unique device identifier, 15 last R-R intervals, and the start time of the recording.

For temporary accumulation and preprocessing of data, a StressMonitor application for Android-operating smartphone is used. Further the data is transmitted via GSM channels to a special server (see Figure 4). The star-shaped topology of the sensor network ensures the efficient use of hardware in case of collective monitoring.

The StressMonitor application displays real-time RR-graphs with spectral frequency-related HRV indices [10] (see Figure 5). It was developed with the use of MySQL 5.5 database, Python 2.7 programming language, Django 1.8 development framework, the Flot 8.2 library, a framework

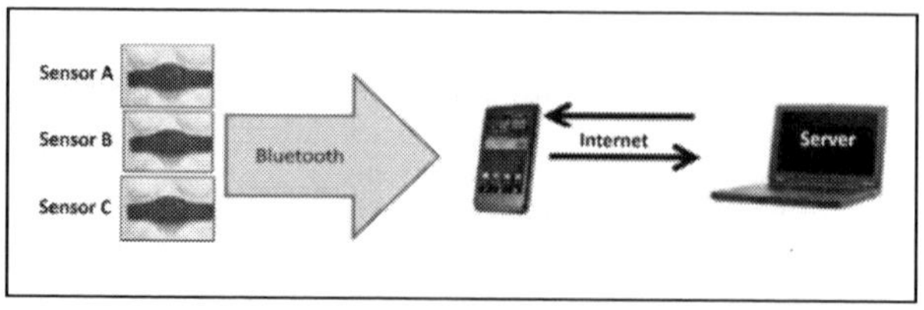

Figure 4. Architecture for wireless registration of HRV in a group of volunteers.

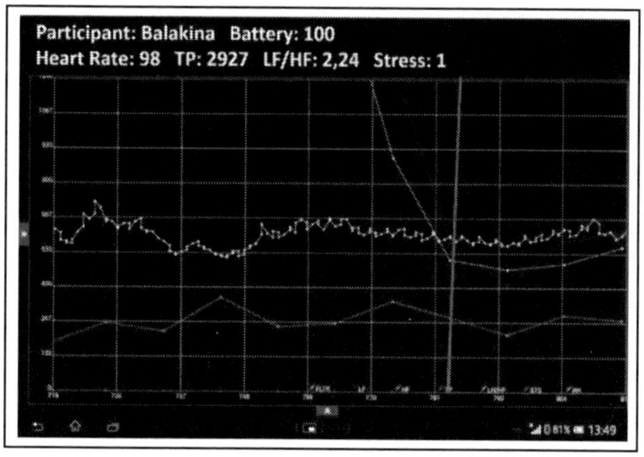

Figure 5. Graphical representation of RR-intervals with calculated HRV spectral indices; the red vertical bar indicates the onset of the stress episode.

for creating a BootStrap 3.2 site design, NumPy, JQuery 2.1 library, Yandex API Maps.

The server database works on MySQL 5.5 software. The received RR signal is cut by a time window for 100 seconds with a time shift of 10 seconds. For the resulting windows, the discrete Fourier transform for uneven signals calculates the spectrum. For the purpose of analysis the spectrum is divided into the following ranges: VLF (0.003-0.04 Hz), LF (0.04-0.15 Hz), HF (0.15-0.4 Hz). The algorithm also uses frequency-related HRV indices: total power (TP) which is a sum of VLF, LF, and HF; vagally-sympathetic balance (LF/HF). As a result, we receive spectral HRV values, GSM coordinates, the time and events associated with stress in a particular person during real-life activity [10].

C. Mathematical Methods for HRV Analysis

For the data processing, spectral periodogram method and statistical methods for analysis of HRV, as well as the continuous wavelet transformation were used.

Spectral analysis is one of the most important types of time series analysis, which allows to determine the effect of autonomic regulation on HRV.

When constructing the spectrum of the rhythmogram, it is important to take into account that the signal itself is not a time series of amplitudes of a physical quantity, but a number series of time intervals between adjacent QRS events in a cardiogram. Fundamentally, the Fourier transform cannot be performed because of the unevenness of the QRS events time scale. It is necessary either to convert a number of intervals into a time series, or to adapt a transformation for a non-uniform series.

Taking into account the specificity of the cardiac signal obtained in real-life contexts, namely, the presence of a nonstationary property and a large number of transition regions, a set of specialized spectral methods for signal processing is suggested:

- Continuous wavelet transform (CW method, or Morlet wavelet) for analysis of amplitude modulations of RR-intervals and spectral components of rhythmograms.
- Dynamic spectral analysis, which synthesizes Fast Fourier transformation algorithms and Lomba-Scargla periodograms, for analysis of rapid changes in the structure of HR. In our case dynamic spectral analysis was performed in a specialized program built in the LabVIEW environment.

As a result, we analyze the temporal dynamics of the power characteristics of the vibration spectra of the RR-intervals, namely: the total power of the HRV spectrum (TP, ms^2); LF, ms^2; HF, ms^2; the ratio of the power of the rhythmogram spectrum (LF/HF).

D. Solution Validation

The validity of the cardiac telemetry complex has been tested by the simultaneous recording with the two certified ECG clinical solutions: AIP Poly-Spectrum (Neurosoft) and Ankar-131 (Medicom) (see Figure 6), and statistical analysis of the HRV patterns and the time sequences of R-R intervals. There is a high correspondence of the time sequences for R-R intervals and the HRV patterns ($p = 0.995$, Student's test).

IV. Results and Discussion

In this section, our selected studies are considered. All the subjects gave informed written consent to participate in the research. Trials with deviation over 10% were removed from the data prior to the analyses. R-R intervals of less than 300 or more than 1500 ms were considered as deviations. Overall, data loss was below 3% across all measurements for each participant group. For all other trials, median filtering was performed prior to spectral analysis.

A. Participants

362 healthy people from Nizhny Novgorod (179 men and 183 women, with a mean age of 21.5 (±1.3)) took part in the experiment. All participants had no heart condition. All of the participants gave an informed consent to stimulation by electrical impulses. All of the participants were naïve regarding the purpose of the experiment.

B. Research Contexts

The data for the research was gathered both in laboratory contexts and in real-life scenarios.

The laboratory contexts were:

- Students during public speaking.
- Students playing a race computer game.
- Sportsmen playing chess with a computer.

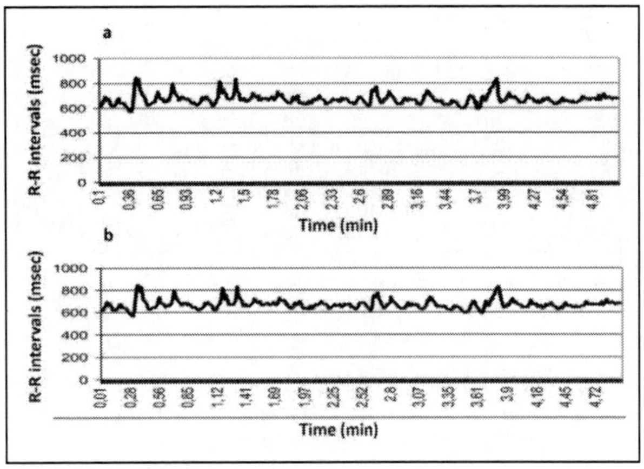

Figure 6. Comparison of synchronized R-R records: (a) Cardiac Telemetry Solution; (b) Poly-Spectrum Analyzer.

The real-life contexts were:

- Drivers of public transport during a sudden maneuver.
- Athletes during an intensive training.
- Firefighters during training in the gas-smoke chamber.

C. Procedure

Volunteers underwent research both individually and collectively, depending on the context.

Before the test, a participant was asked to wear ZephyrTM chest belt. The person responsible for the experiment ran the SmartMonitor app and entered the participant's name, age, and gender. Then the participant was instructed to start the experimentation activity.

Research contexts could take from 30 minutes to several days depending on an activity. All this time, the volunteers wore ZephyrTM chest belt and their data was transferred to the cloud.

After the tests, the lab technician downloaded data from the cloud. HRV spectral analysis was carried out automatically directly in the cloud with the use of a special program complex. Statistical processing of data within each context and between contexts was carried out manually in the program STATISTICA 10.

D. Results

When analyzing spectral HRV indices during public speaking, it was revealed that a decrease in TP with simultaneous increase in LF/HF is a typical pattern for the context (76% of cases).

In the context of computer race game, the moments of mismatch and errors were also characterized by a decrease in TP and an increase in LF/HF values.

During the game of chess in 77% of cases there was a decrease in TP and an increase in LF/HF at the time of a loss (see Figure 7).

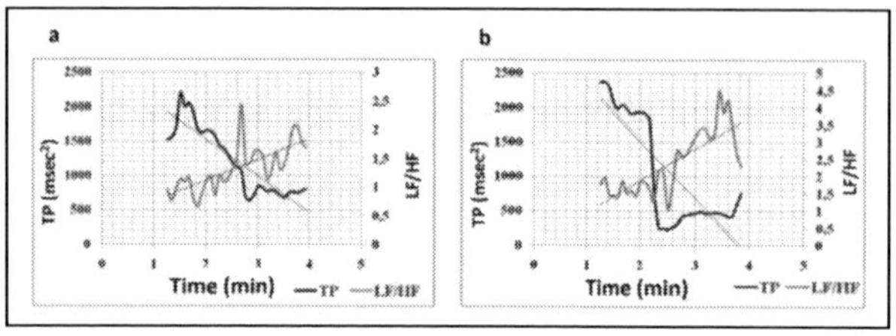

Figure 7. TP and LF/HF HRV spectral values during a game of chess: (a) the player attacks with a subsequent defeat; (b) the player loses the initiative and the game ends in defeat.

In public transport drivers, a sudden maneuver was accompanied by an increase in TP and LF/HF, followed by a decrease in TP with an increase in LF/HF.

HRV spectral analysis in the process of intensive training showed a statistically significant decrease of TP, LF, and HF values after the training load, as well as an increase in LF/HF ($p<0,05$) (see Figure 8). Table IV represents a comparison of the HRV spectral values for warm-up, where optimal HRV values for effective training are identified (see also Figure 9).

TABLE IV. THE OPTIMAL SPECTRAL HRV PARAMETERS DURING WARM-UP

HRV Indices	Mean Value	Standard Error of the Mean
HR, beats per minute	78,40	± 4,53
TP, msec2	3653,82	± 211,81
LF, msec2	1604,49	± 87,59
HF, msec2	573,29	± 61,01
LF/HF	3,09	± 0,86

When analyzing the spectral HRV dynamics in firefighters during training in a gas-smoke chamber, a characteristic R-R pattern was also found: the decrease of TP with an increase in LF/HF (97% of cases).

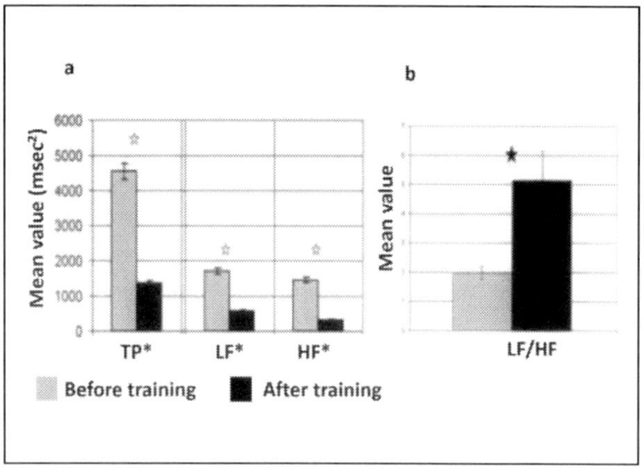

Figure 8. Mean values of spectral HRV parameters before and after exercise: (a) TP, LF, HF (msec2); (b) LF/HF (*p <0.05, Student's test).

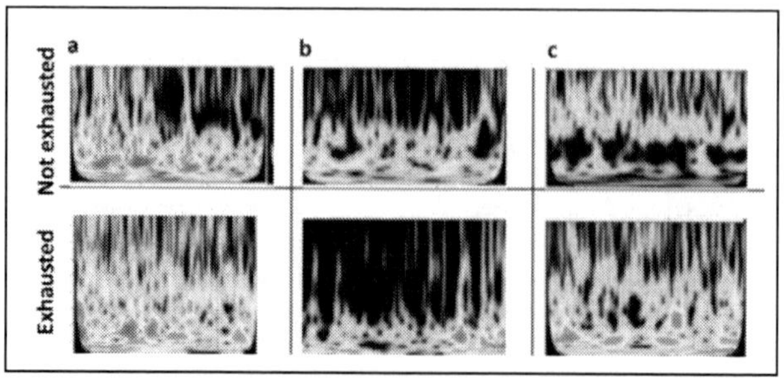

Figure 9. Wavelet analysis of rhythmograms (Morlet wavelets) in athletes at different stages of training: (a) warm-up; (b) training process; (c) rest.

V. CONCLUSION AND FUTURE WORK

To summarize, we undertook an investigation of HRV dynamics during stress in various laboratory contexts and real-life scenarios with the use of cardiac telemetry. We found evidence that there is a dynamic pattern of RR-intervals characteristic for stress, which is the decrease of TP values with the increase of LF/HF values. Therefore, stress is physiologically manifested by increased sympathetic and diminished parasympathetic activity of the ANS.

The main objective of the study was to analyze the continuous dynamics of autonomic regulation of HRV in heterogeneous laboratory and real-scenarios and to identify the dynamic structures of autonomic regulation indices specific for the stress response.

The results of the study of the HRV autonomic regulation dynamics under variousloads reproduce and confirm data on the high sensitivity of the regulation system to various changes in the internal and external contexts of a living system [25].

Using the new wireless cardiac telemetry solution that was developed specifically for the task of measuring living systems in the real-life scenarios opened new opportunities for research [51-52].

As the result of experimental series, it is shown that the dynamics of the HRV autonomic regulation parameters is nonlinear, quasi-periodic, and does not have direct relationships with the autonomous regulation measurements in stationary contexts like sitting or lying. There are no grounds to deny the possibility of revealing the dependencies between the indices of the autonomic regulation at rest and in the context of natural activity. However, they seem to be individual. Therefore, an appropriate experimental design presupposes repeated long-term monitoring of a person in a rich set of real-life scenarios.

Use of spectral analysis algorithms for uneven time series, such as the sequence of RR-intervals, made it possible to reveal VHF spectrum of HRV that was not considered in classical physiology and medicine.

Mathematical methods of dynamic spectral analysis allow scaling the discreteness of the received parameters, so there is an opportunity to observe real-time changes in the structure of the autonomous regulation of HR in the course of real-life events. This makes it possible to use 15-second rhythmograms for analysis instead of 300-second rhythmograms as in the classical HRV analysis.

The novelty of the study is the approach with the loads differing not in the complexity, but in the very nature of the target task.

In general, the decrease in HRV as a sign of the presence of a disease is discussed in many works [58-60]. The list of diseases is very wide, including cardiovascular system disorders (myocardial dysfunction, tetraplegia, hypertension, congestive circulatory insufficiency, chronic mitral regurgitation, cardiomyopathy, ventricular arrhythmias, supraventricular arrhythmias, etc.); psychoneurological disorders (posttraumatic stress disorder, depression, anxiety), oncological diseases, infectious diseases (influenza, ARVI). Apparently, the decrease in HRV is a nonspecific marker of the presence of a disease. Then, it is debatable that the famous Selye triad, which is a complex of three stress-specific symptoms (hypertrophy of the adrenal glands, thymus involution, ulceration in the digestive tract), can be extended. In order to introduce the decrease in HRV as the dynamic marker of stress, we need to reproduce pharmaco-physiological responses associated with stress reaction.

The HRV data obtained in athletes during physical load represents a fundamentally new context for research of stress. An important result was obtained when trying to reveal the coordination between the initial states of athletes with their performance in the context of uninterrupted monitoring of their training activities from warm-up until rest.

According to the 3-component theory of stress [8] and empirical data, we assume that the first phase of stress-related HRV pattern dynamics is associated with the activation of all regulatory loops, and especially the sympathetic component of the autonomic nervous system. The second phase of HRV dynamics may be associated with the activation of the endogenous opioid system (EOS) (see Figure 9).

A contradiction in interpretation may arise when comparing the two indices of the sympathetic activity, which is the power of the LF component of HRV and HR spectrum. It is commonly believed that an increase in HR and an increase in the LF index reflect the activity of the sympathetic system. In our case, the increase in HR values is combined with the decrease of the LF component (i.e., training load effects in athletes). That is, the two indices contradict each other.

However, experimental data on the LF component dynamics were historically obtained in static contexts, that is, active contexts were not considered. At rest, there is a direct correlation between the power of LF and HR.

As a result, changes in HR and LF power are not always associated with the activity of the sympathetic subsystem. Apparently, they reflect different regimes of sympathetic influence on the heart rhythm. The LF component reflects the phasic modulation of the HR in the context of rest. The increase in HR is a single strong increase associated with sympathetic activation in the dynamic real-life context.

The results of the use of the cardiac telemetry in numerous experiments justify its potential for application in further research into stress phenomenon, health promotion

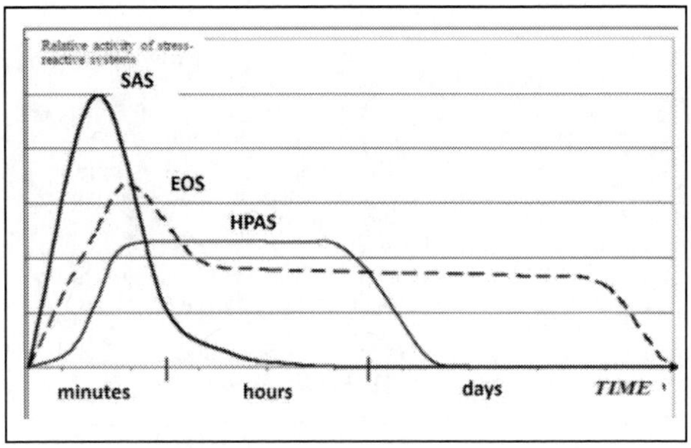

Figure 10. The dynamics of the stress-protection systems SAS, HPAS, and EOS during stress.

services, consumer wellness devices, professional sports, and amateur training.

In future researches we would like to focus on defining the best practices for stress management. Research into neurobiological mechanisms of stress at the cell level is planned for further technology development.

ACKNOWLEDGMENT

This work was supported by the Russian Foundation for Humanities under grant 15-06-10894_a, grant 16-06-00133_a, and RFBR, grant 16-06-00501_a.

REFERENCES

[1] S.B. Parin, S.A. Polevaia, and A.V. Polevaia, "A Neurochemical Framework to Stress and the Role of the Endogenous Opioid System in the Control of Heart Rate Variability for Cognitive Load," in COGNITIVE 2017: 9th International Conference on Advanced Cognitive Technologies and Applications, IARIA, pp. 16–20, 2017.

[2] E-M. Backe, A. Seidler, U. Latza, K. Rossnagel, and B. Schumann, "The role of psychosocial stress at work for the development of cardiovascular diseases: a systematic review," International Archieves of Occupational and Environtal Health, vol. 85, pp. 67–79, 2012.

[3] S. C. Segerström and G. E. Miller, "Psychological Stress and the Human Immune System: A Meta-Analytic Study of 30 Years of Inquiry," Psychological Bulletin, vol. 130(4), pp. 601–630, 2004.

[4] N. Schneiderman, G. Ironson, and S. D. Siegel "Stress and Health: Psychological, Behavioral, and Biological Determinants," Annual Review of Clinical Psychology, vol. 1, pp. 607-628, 2005.

[5] World Health Organization, "Charting the way to well-being," The European Health Report, 2012.

[6] A. Muaremi, B. Arnrich, and G. Troster, "Towards Measuring Stress with Smartphones and Wearable Devices During Workday and Sleep," Bionanoscience, vol. 3, pp. 172–183, May 2013, doi:10.1007/s12668-013-0089-2.

[7] H. Selye, "The story of the adaptation syndrome," Montreal: ACTA, Inc., Medical Publishers, 1952.

[8] S. B. Parin, "Humans and animals in emergency situations: neurochemical mechanisms, evolutionary aspect," Vestnik NSU, vol. 2(2), pp. 118–135, 2008.

[9] S. B. Parin and S. A. Polevaya, "Features of information transformation under stress and shock," VIII All-Russian

Scientific and Technical Conference "Neuroinformatics-2006", Moscow: MEPhI, pp. 165-171, 2006.

[10] P. K. Anokhin, "Essays on the Physiology of Functional Systems," Moscow: Medicine, pp. 17-59, 1975.

[11] S. A. Polevaya, S. B. Parin, and E. G. Stromkova, "Psychophysical mapping of human functional states," Experimental Psychology in Russia: Traditions and Perspectives, pp. 534-538, 2010.

[12] R. M. Bayevsky, "Analysis of heart rate variability: history and philosophy, theory and practice," Clinical informatics and telemedicine, vol. 1(1), pp. 54-64, 2004.

[13] G. E. Billman, "Heart rate variability–a historical perspective," 2011, doi:10.3389/fphys.2011.00086.

[14] A. S. Ban, N. A. Paramonova, G. M. Zagorodnyi, and D. S. Ban, "Analysis of the interrelation between heart rate variability indices," Military medicine: a scientific and practical peer-reviewed journal, vol. 4, pp. 21-24, 2010.

[15] V. A. Barabanschikov, "Experimental Psychology in Russia," Moscow: Institute of Psychology RAS, pp. 13-18, 2010.

[16] H. Bonnemeier, G. Richardt, J. Potratz, U. K. Wiegand, A. Brandes, N. Kluge, and H. A. Katus, "Circadian profile of cardiac autonomic nervous modulation in healthy subjects: differing effects of aging and gender on heart rate variability," J. Cardiovasc. Electrophysiol, vol. 14(8), pp. 791-799, 2003.

[17] G. B. Stanley, K. Poolla, and R. A. Siegel, "Threshold modeling of autonomic control of heart rate variability," IEEE Trans Biomed Eng, vol. 47(9), pp. 1147-1153, 2000.

[18] A. Zaza and F. Lombardi, "Autonomic indexes based on the analysis of heart rate variability: a view from the sinus node," Cardiovasc Res, vol., 50(3), pp. 434-442, 2001.

[19] R. M Baevsky and G. G. Ivanov, "Heart rate variability: theoretical aspects and possibilities of clinical application," Ultrasound functional diagnostics, vol. 3, pp. 108-127, 2001.

[20] D. Singh, K. Vinod, and S. C. Saxena, "Sampling frequency of the RR interval time series for spectral analysis of heart rate variability," J. Med. Eng. Technol, vol. 26(6), pp. 263-272, 2004.

[21] E. Toledo, I. Pinhas, D. Aravot, and S. Akselrod, "Very high frequency oscillations in the heart rate and blood pressure of heart transplant patients," Med. Biol. Eng. Comput, vol. 41, pp. 432-438, 2003.

[22] S. V. Bozhokin, E. M. Lecova, V. O. Samoilov, and P. I. Tolkachev, "Wavelet analysis of non-stationary heart rate variability during passive orthostatic test," Biophysics, vol., 4, pp. 696-712, 2012.

[23] C. Winkler, "Attention and respiration," Proc. Acad. Sci. Amsterdam, vol. 1, pp. 121-138, 1899.

[24] V. A. Mashin, "Three-factor model of heart rate variability in psychological research of functional systems of human operator," Moscow, 2010.

[25] V. V. Parin and F. Z. Meerson, "Essays on the clinical physiology of blood circulation," Moscow: Medicine, 1960.

[26] J. Taelman, S. Vandeput, E. Vlemincx, A. Spaepen, and S. Van Huffel, "Instantaneous changes in heart rate regulation due to mental load in simulated office work," Eur J Appl Physiol, vol. 111(7), pp. 1497-1505, 2011a.

[27] J. Taelman, U. Katholieke, L. B. Leuven, S. Vandeput, I. Gligorijevic, and A. Spaepen, "Timefrequency heart rate variability characteristics of young adults during physical, mental and combined stress in laboratory environment," Conf Proc IEEE Eng Med Biol Soc, pp. 1973-1976, 2011b.

[28] E. K. S. Hanson, G. L. R. Godaert, C. J. M. Maas, and T. F. Meijman, "Vagal cardiac control throughout the day: the relative importance

[29] of effort-reward imbalance and within-day measurements of mood, demand and satisfaction," Biol. Psychol, vol. 56, pp. 23-44, 2001.

[30] A. Collins, A. Karasek, and K. Costas, "Job strain and autonomic indices of cardiovascular disease risk," American Journal of Industrial Medicine, vol. 48, pp. 183-193, 2005.

[31] P. Melillo, M. Bracale, and L. Pecchia, "Nonlinear Heart Rate Variability features for real-life stress detection. Case study: students under stress due to university examination," Biomed. Eng.Online, vol. 10(1), pp. 96, 2011.

[32] T. W. Buchanan and D. Tranel, "Stress and emotional memory retrieval: effects of sex and cortisol response," Neurobiol. Learn. Mem, vol. 89, pp. 134-141, 2008.

[33] R. Nikolova, L. Tzaneva, and S. Danev, "Investigation of noise exposure effect on heart rate variability parameters," Cent Eur. J. Public Health, vol. 9(3), pp. 130-132, 2001.

[34] T. Chandola, A. Britton, E. Brunner, H. Hemingway, M. Malik, M. Kumari, E. Badrick, M. Kivimaki, and M. Marmot, "Work stress and coronary heart disease: what are the mechanisms? ," Eur Heart J, vol. 29, pp. 640-648, 2008.

[35] T. Chandola, A. Heraclides, and M. Kumari, "Psychophysiological biomarkers of workplace stressors," Neurosci. Biobehav. Rev, vol. 35(1), pp. 51-57, 2010.

[36] N. Rohleder, J. M. Wolf, E. F. Maldonado, and C. Kirschbaum, "The psychosocial stress-induced increase in salivary alpha-amylase is independent of saliva flow rate," Psychophysiology, vol. 43, pp. 645-652, 2006.

[37] M. Altemus, L. S. Redwine, Y. Leong, C. A. Frye, S. W. Porges, and C. S. Carter, "Responses to laboratory psychosocial stress in postpartum women," Psychosom. Med, vol. 63, pp. 814-821, 2001.

[38] H. Lackschewitz, G. Hüther, and B. Kröner-Herwig, "Physiological and psychological stress responses in adults with attention-deficit/hyperactivity disorder (ADHD)," Psychoneuroendocrinology, vol. 33, pp. 612-624, 2008.

[39] M. Malik, J. T. Bigger, A. J. Camm, R. E. Kleiger, A. Malliani, A. J. Moss, and P. J. Schwartz, "Heart rate variability: Standards of measurement, physiological interpretation, and clinical use," Eur. Heart J, vol. 17(3), pp. 354-381, 1996.

[40] R. M. Baevsky, "Prediction of states on the verge of norm and pathology," Moscow: Medicine, 1979.

[41] P.K.. Anokhin, "Fundamental questions of the general theory of functional systems of the principles of the system organization of functions", Moscow: Science, pp. 5-61, 1973.

[42] J. R. Turner, D. Carroll, S. Dean, and M. G. Harris, "Heart rate reactions to standard laboratory challenges and a naturalistic stressor," International Journal of Psychophysiology, vol. 151, pp. 151-152, 1987.

[43] C. Bernard, "The physiology of the heart and its relation to the brain: lecture," N. Solovyov, Ed. St. Petersburg, 1867.

[44] J.F. Thayer, F. Åhs, Fredrikson, M.J.J Sollers, and T.D. Wagere, "A meta-analysis of heart rate variability and neuroimaging studies: Implications for heart rate variability as a marker of stress and health," Neuroscience and Biobehavioral Reviews, 36, pp. 747–756, 2012.

[45] J.F. Thayer and R.D. Lane, "A model of neurovisceral integration in emotion regulation and dysregulation," J. Affect. Disord., 61, pp.201–216, 2000.

[46] J.F. Thayer and R.D. Lane, "Claude Bernard and the heart–brain connection: further elaboration of a model of neurovisceral integration," Neurosci. Biobehav. Rev, 33, pp.81–88, 2009.

[47] Y. J. Zanstraa and D. W. Johnston, "Cardiovascular reactivity in real life settings: Measurement, mechanisms and meaning," Biol. Psychol, vol. 86(2), pp. 98-105, 2011.

[48] C. Kirschbaum, K. M. Pirke, and D. H. Hellhammer, "The 'Trier Social Stress Test'—a tool for investigating psychobiological stress responses in a laboratory setting," Neuropsychobiology, vol. 28, pp. 76-81, 1993.

[49] J. R. Turner, S. S. Girdler, A. Sherwood, and K. C. Light, "Cardiovascular responses to behavioral stressors: laboratory-field generalization and inter-task consistency," Journal of Psychosomatic Research, vol. 34(5), pp. 581-589, 1990.

[50] L. Warwick-Evans, J. Walker, and J. A. Evans, "A comparison of psychologically-induced cardiovascular reactivity in laboratory and natural environments," Journal of Psychosomatic Research, vol. 32, pp. 493-504, 1988.

[51] L. J. Van Doornen and R. W. Van Blokland, "The relation to type a behavior and vital exhaustion with physiological reactions to real life stress," Journal of Psychosomatic Research, vol. 3, pp. 715-725, 1989.

[52] J. P. Davig, K. T. Larkin, and J. L. Goodie, "Does cardiovascular reactivity to stress measured in the laboratory generalize to thesis and dissertation meetings among doctoral students? ," International Journal of Behavioral Medicine, vol. 7, pp. 216-235, 2000.

[53] E.V. Runova, S.B. Parin, M.M. Nekrasova, A.V. Bakhchina, A.V. Kovalchuk, I.S. Shyshalov, and S.A. Polevaya, "Monitoring and distant diagnostics of sportsment's functional state based on information technologies and telemetry in the conditions of natural activity," Int J Psychophysiology, vol. 85(3), pp. 420– 421, 2012.

[54] S. A. Polevaia, S. B. Parin, A. V. Bakhchina, M. M. Nekrasova, I. S. Shishalov, E. V. Runova, and V. V. Kozhevnikov, "Determination system of functional state of a group of people," Patent RF №129680, 2013.

[55] S. A. Polevaya, S. B. Parin, A. V. Bakhchina, M. M. Nekrasova, I. S. Shishalov, E. V. Runova, and V. V. Kozhevnikov, "Determination system of functional state of a group of people with feedback," Patent RF №129681, 2013.

[56] N. I. Shlyk, A. E. Alabuzhev, N. Z. Feofilaktov, and A. O. Sleptsov, "Dynamic studies of heart rate variability in mid-season athletes in the training process in mid-mountain conditions," Materials of the 5th All-Russian Symposium with international participation: Variability of the heart rhythm. Theoretical aspects and practical application, pp. 384-389, 2011.

[57] H. Che-Hao, T. Ming-Ya, H. Go-Shine, L. Tso-Chou, C. Kuen-Pao, H. Shung-Tai, S. LiangYu, and L. Chi-Yuan, "Poincaré plot indexes of heart rate variability detect dynamic autonomic modulation during general anesthesia induction," Acta Anaesthesiologica Taiwanica, vol. 50, pp. 12-18, 2012.

[58] K. Becker, I. Görlach, and T. Frieling, "Häussinger, D. Characterization and natural course of cardiac autonomic nervous dysfunction in HIV-infected patients," AIDS, vol., 11(6), pp. 751-757, 1997.

[59] S. Cohen and N. Weinstein, "Nonauditory effects of noise on behavior and health," Journal of Social Issues, vol. 37, pp. 36-70, 1981.

[60] M. De Couck, D. van Brummelen, D. Schallier, J. De Grève, and Y. Gidron, "The relationship between vagal nerve activity and clinical outcomes in prostate and non-small cell lung cancerpatients," Oncology reports, pp. 2435-2441, doi:10.3892/or.2013.2725.

[61] N. Egelund, "Spectral analysis of heart rate variability as an indicator of driver fatigue," Ergonomics, vol. 25(7), pp. 663-672,1982.

[62] A.Y. Kaplan, "Heart rate variability and the nature of feedback upon the result of operational activity in a human," Zh Vyssh Nerv Deiat I P Pavlova, vol.48(6), pp.345-350, 1999.

[63] G. Mulder, L.J.M. Mulder, T.F. Meijman, J.B.P. Veldman, and A.M. Van Roon, "A psychophysiological approach to working conditions," in Engineering Psychophysiology: Issues and Applications, R.W. Backs, W. Boucsein, Eds. pp. 139-159,2000.

[64] W.P. Brinkman, R. Haakma, and D.G. Bouwhuis, "Memory Load: a Factor that Links the Usability of Individual Interaction Components Together," Proc. the Conference HCI: Design for Life, vol.2, Leeds, UK, pp.165-168, 2004.

Experimentally Analyzing Relationships between Learner's Status in the Skill Acquisition Process and Physiological Indices

Yoshimasa Ohmoto
Department of Intelligence Science and Technology
Graduate School of Informatics
Kyoto University
Kyoto, Japan
Email: ohmoto@i.kyoto-u.ac.jp

Takahiro Matsuda
Department of Intelligence Science and Technology
Graduate School of Informatics
Kyoto University
Kyoto, Japan
Email: matsuda@ii.ist.i.kyoto-u.ac.jp

Toyoaki Nishida
Department of Intelligence Science and Technology
Graduate School of Informatics
Kyoto University
Kyoto, Japan
Email: nishida@i.kyoto-u.ac.jp

***Abstract*—There are many tasks for which people need domain-specific skills learned through long-term practice. Many skill acquisition models have been proposed that include the mental states of the learners, but only few studies have tried to estimate mental states using objectively measured data. The purpose of this study was to experimentally investigate whether the status of the skill acquisition process could be estimated by physiological indices that indicate the learner's mental states. For this purpose, we conducted an experiment to obtain data on physiological indices and a subjective report of the feeling of difficulty during the skill acquisition task. As a result, we confirmed the relationship between the participant's statuses and the physiological indices. In addition, we classified the trials with the feeling of difficulty in the experiment using Support Vector Machine. As a result, the values of accuracy were over 0.65 when the data not used to calculate the SVM model were classified. We could show that the physiological indices are helpful clues to estimate the status of the skill acquisition process.**

Keywords–Skill acquisition model; Mental state estimation; Physiological indices.

I. INTRODUCTION

This article is a substantially extended version of the authors' paper "Experimentally Analyzing the Skill Acquisition Model using Task Performance and Physiological Indices" [1], presented at COGNITIVE 2017, the Ninth International Conference on Advanced Cognitive Technologies and Applications, IARIA in Greece.

A skill is the ability to perform a task with pre-determined results, often involving a given amount of time, energy, or both. There are many tasks for which people need domain-specific skills learned through long-term practice. We call this kind of undertaking a "skill task." To acquire the abilities for a skill task, people usually train by carrying it out repeatedly. However, it is difficult to learn the aptitude in question alone because, in many cases, people cannot objectively monitor their skill level and task performance. Experts and instructors can support learners, but the methods to support learning a skill task have been established for few endeavors. Some systems have been proposed to support learning based on task performance.

Generally, when a person has acquired a skill, i.e., proficiency or expertise, this means that person can efficiently carry out a specific task. Previous studies, e.g., [2], are founded on the principle that learners pass through five stages: Novice, competence, proficiency, expertise, and mastery. The phases are characterized by how rules interplay with real-world context. A novice will simply follow the rules that they are given and not consider context. Intermediate stages contain a mix of rule following, combined with more and more sophisticated consideration of context [3]. To assess skill level, the skill task is divided into a number of sub-tasks, the performance on which is rated. However, to competently perform a job, many skill tasks require various sub-skills corresponding to the sub-tasks. In this case, even when the learner has acquired some sub-skills, a situation may arise where he/she cannot synthetically use them. In addition, it is often difficult to rate skill task performance itself. Tranquillo and Stecker [3] pointed out that a master may, in fact, not be able to state the rules or the heuristics that they are using. For example, we cannot rate presentation skills independent of the contents of the presentation. To support the skill acquisition of the skill task, identifying the statuses in the skill acquisition process, e.g., the steep acceleration section or the plateau section in a learning curve, is more useful than rating the skill level. To identify this status, human instructors focus on the learner's responses to unknown situations (such as questions and answers), giving a new challenge, and deliberately making mistakes. In other words, they evaluate the skill task model that the learner has. Of course, the skill task model cannot be observed directly.

The important point is that we can use the learner's own recognition of his/her skill task model. The Dreyfus model of skill acquisition [2] shows how students acquire abilities through formal instruction in addition to practicing. This model identifies skill level based on four binary qualities: (1) recollection (non-situational or situational); (2) recognition (decomposed or holistic); (3) decisions (analytical or intuitive); and (4) awareness (monitoring or being absorbed). The model is intuitive, but no one knows how to evaluate the four factors concretely. We considered an approach to estimate the learner's recognition, i.e., the learner's mental state, from objectively measured data during the skill acquisition process. For this assessment, we used physiological indices that could gauge the responses of the learner's autonomic nervous system related to his/her mental states. The physiological indices are usually employed to ascertain mental stress.

Kraiger et al. [4] attempted to move toward a training evaluation model by developing a classification scheme for evaluating learning outcomes. They integrated theory and research from a number of diverse disciplines and provided a multidimensional perspective on learning outcomes. They provided a classification scheme for learning outcomes for training evaluation but were unable to propose a method to identify the outcomes objectively. Mitchell et al. [5] assessed participants' self-efficacy goals, expected performance, and the degree to which certain judgments required cognitive processing. The results showed that self-efficacy was a better predictor of performance than were expected score or goals on early trials, whereas the reverse was true for later trials. This result showed that the mental state is useful for estimating the skill level because the responses changed with the skill acquisition. From these results, we think that we needed a method to objectively estimate the learner's mental state that changed with the skill acquisition process.

The final goal of this study was to develop a method to determine the status of the skill acquisition process in detail for the skill task, using task performance and the learner's mental states. For this purpose, we experimentally investigated whether the status of the skill acquisition process could be estimated by the physiological indices that can be used to estimate the learner's mental states. We conducted an experiment to obtain the data of the physiological indices and a subjective report of the learner's mental states.

The paper is organized as follows. Section II briefly introduces previous works on the skill acquisition model and assessing human stress. Section III explains the outline of the technique for appraising human stress and the skill acquisition model via two dimensions: (1) task performance and (2) the learner's mental states. Section IV describes the experiment and the analysis of the data. Section V discusses the analysis, and Section VI lays out our conclusions and future works.

II. RELATED WORK

The Dreyfus Model of skill acquisition [2] seems to be one of the most famous models of how students acquire skills through formal instruction and practicing. This model is used in a broad range of tasks, such as defining an appropriate level of competence, supporting judgements of when a learner is ready to teach others, and so on. However, there are some criticisms of this model [6]. According to these authors, there is no empirical evidence for the presence of stages in the development of expertise.

Day et al. [7] examined the viability of knowledge structures as an operationalization of learning in the context of a task that required a high degree of skill using a complex video game. At the end of acquisition, participants' knowledge structures were assessed and the similarity of trainees' knowledge structures to an expert structure was correlated with skill acquisition and was predictive of skill retention and skill transfer. In addition, knowledge structures mediated the relationship between general cognitive ability and skill-based performance. However, they did not mention the process of change of the cognitive ability during the skill acquisition.

Langan-Fox et al. [8] summarized traditional models ([9][10][11]). They pointed out that models of skill acquisition largely ignore the experiences and dynamic internal processes of a person while learning a skill. They attempted to highlight the importance of a dynamic description of skill acquisition in their research. Process-oriented factors such as motivation, memory, interruptions, emotion, and metacognition are investigated in relation to skilled performance. However, their discussions were conceptual and they did not investigate the matter experimentally.

There are also some studies focusing on the learning process, e.g., [12][13][14]. This is called the learning curve, and it is well known that skills are acquired gradually, followed by a period of steep acceleration and then a plateau phase. However, there is no definite answer as to why such stepwise learning occurs. In general, it is thought that in the learning status where changes in cognition for tasks are required, learning is in the plateau section until the change occurs. However, it is difficult to objectively observe how the change occurs. Observing changes in mental state during the skill acquisition is also important in clarifying such processes.

There are some studies in which the physiological indices were used for identifying the mental states. Patterson et al. [15] used heart rate variability (low frequency:high frequency ratio) to differentiate invested cognitive effort during the acquisition and retention of a novel task. They found the usefulness of heart rate variability in discriminating the cognitive effort invested for a recently acquired skill. Walker et al. [16] sought to identify a physiological measure that could help predict team performance during a complex and dynamic task. Regression analyses showed that team autonomic activity accounted for 10% of the variance in team performance scores. They showed that the task performance can be predicted from physiological indices. Parsinejad [17] tried to infer the mental workload changes of human operator using physiological measurements and performance metrics, while keeping participants' inexperience as the key parameter. This research showed that physiological indices and the participant's finger-stroke patterns on the touch screen could help flag unfamiliarity of participants in the difficult game. These studies showed that the physiological indices were useful to identifying the learner's mental states and they could predict the task performance based on the physiological indices and the learner's behavior. However, they did not focused the relationships between the mental states and the learning process.

III. THE SKILL ACQUISITION MODEL, INCLUDING MENTAL STATES

Some previous studies have proposed the skill acquisition model, which considers learners' mental states [2][8]. However, the mental states were evaluated by human observation. In other words, previous studies have not focused on how to measure, evaluate, and use mental states via objective approaches to the skill acquisition model. In addition, they have centered on mental states through the lens of a specific skill level, but have not concentrated on the dynamics of people going through a learning process. This study aimed to develop a technique for assessing a learner's skill level based on his/her performance and mental states. To do so, we first confirmed the relationship between subjective reports on mental states and the measured physiological indices. We then interpreted the data and performance for estimating skill levels and the process of learning. In this section, we explain the physiological indices that were used to appraise mental states, and propose a skill acquisition model that has two dimensions that define skill level.

A. Physiological indices for assessing mental states

In this study, we propose the skill acquisition model using task performance and mental states. We especially focus on how to gauge and use mental states in the skill acquisition model. It is hard to use a learner's behavior to evaluate mental states because the learner's behavior depends on the task and skill level. However, when a learner feels that an activity is difficult, he/she feels mental stress due to his/her line of thinking and stimuli from the endeavor. Therefore, we consider stress useful for appraising a typical mental state in the learning process.

Many previous investigations have reported on physiological indices for estimating mental stress. However, in ongoing daily interactions, we can often find concurrent physiological responses that are not related to the event. One reason is that people involved in continuous interactions often plan their actions, such as what to tell and how to move. Therefore, to assess human mental states, we had to consider the context of an interaction and the response characteristics of the physiological indices.

Physiological indices are biological reactions caused by the autonomic nervous system, for example: brain waves, potential differences in cardiographs, variations in blood pressure, pulse waves, respiration, body temperature, muscle potential, and skin conductance. In continuous interactions, some of these are susceptible to noise from body motions. We used skin conductance responses (SCR) and electrocardiograms (Low Frequency (LF)/High Frequency (HF) values) because these are relatively resistant to noise.

Since the underlying mechanisms of SCR and electrocardiograms are different, we expected that they could be used to distinguish between different responses from various sources of stress. Sweating is controlled by the sympathetic nervous system [18] and can be elicited by emotional stimuli, intellectual strain, or painful cutaneous stimulation. The underlying mechanisms of SCRs are more related to anticipation, expectation, and attention concentration [19]. We thus anticipated that SCRs could be used to tell when someone is dealing with an unexpected situation.

For electrocardiograms, the LF/HF value is calculated using instantaneous heart rate. It shows heart rate variability (HRV), which is controlled by the sympathetic and parasympathetic nervous systems and humoral factors. The underlying mechanisms of HRV are complex. Lacey and Lacey [20] suggested that it is caused by sensory intake and sensory rejection. In addition, the parasympathetic nervous system responds quickly (< 1 s) to stimuli. We thus thought that the LF/HF (HRV) would show reactive responses based on external stimuli.

B. Relationship between a mental state and skill acquisition

The performance on a skill task is often difficult to rate. We consider skill acquisition to mean that a learner has constructed an efficient task model. If we could directly evaluate the learner's skill task model, we could identify the statuses of the skill acquisition process. Human instructors focus on the learner's responses to unknown situations to estimate the skill task model. The learner's responses are expressions that the learner feels difficulty and senses the patterns of mistakes. Since learners might recognize the feeling of difficulty and the mistakes consciously or unconsciously, we might estimate them based on the learner's mental states.

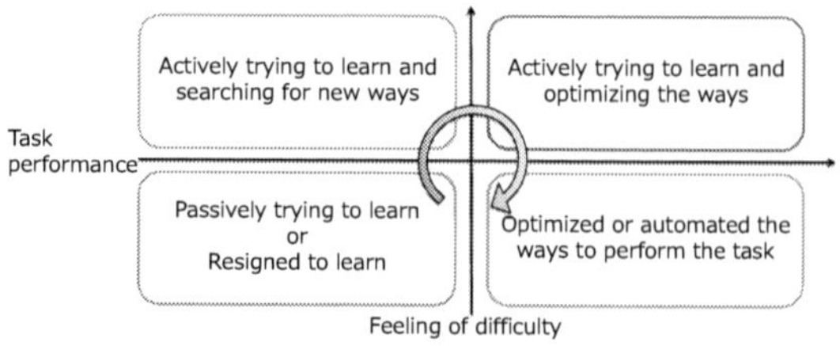

Figure 1. The outline of the skill acquisition model via two dimensions.

We focused on mental state changes to estimate whether the learners thought their constructed task models were efficient. In general skill acquisition processes, we can draw a learning curve whose shape resembles a series of steps, e.g., [12][13][14]. In the steep acceleration section of the curve, the learners receive feedback from subjective and objective evaluations and improve their behavior. In the plateau section, the learners optimize their behavior and sometimes execute trial and error for the next step to improve the task performance. When the learners are constructing the skill task model, they experience intrinsic stresses. When the learners are improving their model, they mainly experience extrinsic stresses. As mentioned above, we expect that the intrinsic and extrinsic stress can be estimated with physiological indices. By using these estimates, we may estimate whether the learners thought their constructed task models were efficient.

C. The skill acquisition model in two dimensions

We propose a skill acquisition model with two dimensions-task performance and mental state-when a person feels that a task is challenging (we call this a "feeling of difficulty"). Figure 1 shows the model and the two dimensions. In the traditional three phase model, Phase 1 corresponds to the lower left part, Phase 2 corresponds to the upper part, and Phase 3 corresponds to the lower right part. The "task performance" of the horizontal axis is the task result, objectively quantified. The "feeling of difficulty" of the vertical axis is a mental state, assessed based on the measured physiological indices. The traditional skill acquisition model assumes that task performance (weakly) rises with an increase in skill levels. When a skill can be segmented into small mutually independent sub-skills, this assumption may be true of the process of acquiring sub-skills. However, attaining sub-skills does not contribute to gaining more than a certain level of overall aptitude for many skill tasks, such as ballroom dancing. In this case, the synthetic use of sub-skills is often important, and indeed itself is a target of skill acquisition. When the learner practices synthetic use through trial and error, task performance often decreases. In our model, the trial and error process is included in the lower part. When the task performance decreases but the difficulty feeling is low, the model interprets the process as trial and error. We expected that the learner would cycle through this model throughout the skill acquisition process.

We assume that the skill acquisition process unfolds in the following manner. The initial state of the learner is in the upper or lower left part, i.e., performance is low. The learner usually needs trial and error to learn the skill task, so the state of the learner is maintained or transitions to the upper part. Through the learning process, task performance increases and

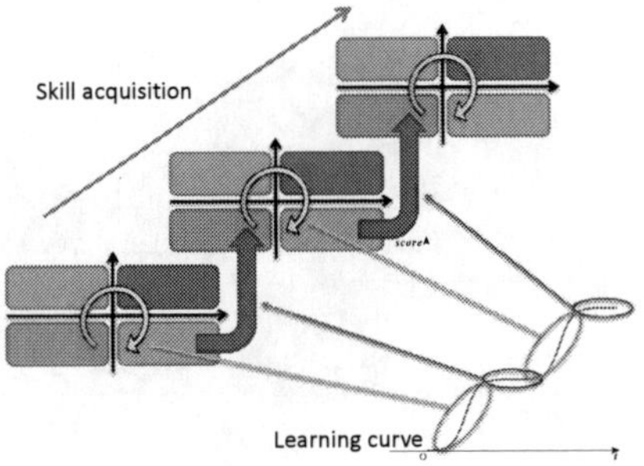

Figure 2. The outline of the skill acquisition process whose shape resembles a series of steps.

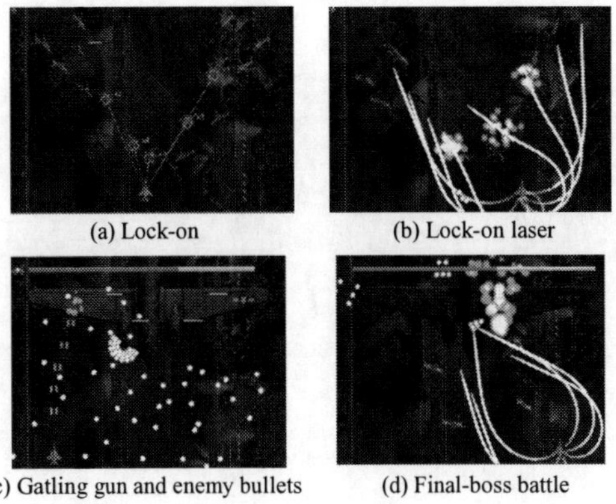

Figure 3. The screen shots of the shooting game.

the learner's state moves to the upper right part. In this state, the learner can perform the task at hand more efficiently than before, but he/she is not accustomed to or does not understand how to carry out the activity. Through practice at this stage, the learner can synthetically and automatically perform the task, and his/her state moves to the lower right part. In a common case, the learner continuously performs trial and error to find better ways, so task performance sometimes falls. In this case, the learner's state is ready for the next stage of the skill acquisition process. If the learner can find clues for better ways to perform the task, the skill acquisition process advances to the next phase. If the learner cannot find better ways, the skill acquisition process terminates in the lower right part of the stage. Figure 2 shows a series that demonstrates the flow of the skill acquisition process in our model. Through the cyclic skill acquisition process, learners deepen their understanding of the skills and the tasks. After that, they can find a breakthrough for better ways to perform the task, and advance to the next phase of the skill acquisition.

IV. Experiment

The purpose of this experiment was to investigate whether the physiological indices were related to the subjective reports about the feeling of difficulty toward the task, and whether we could evaluate the state of transition in the proposed skill acquisition model based on task performance and physiological indices. We adopted a shooter game as a skill task. Some previous studies, e.g., [7], have adopted various shooter games. The shooter game that we developed has features of the skill task. The advantages of using a shooter game as the skill task include the following: (1) The learner needs to obtain game playing skills, which is hard to verbalize; (2) We can control the difficulty of a task; and (3) We can easily analyze the skill acquisition process because we can independently divide the time series of the game events, which is the target of the skill acquisition. In addition, learners can repeatedly play the game with high motivation. We conducted an experiment in which the participants played the shooter game repeatedly, and we obtained the game scores and physiological indices during game play. After the experiment, we analyzed the data to confirm the relationship between the physiological indices and the subjective reports about the feeling of difficulty. Furthermore, we examined the relationship in the skill acquisition process.

A. *Task*

To achieve the best performance in the shooter game, the player must cultivate certain skills and gain certain knowledge, such as operation procedures, scoring rules, the way to defeat one's enemies, the features of game stages, basic survival patterns, and specific techniques to obtain a high score. Some of these cannot be verbalized and the best method varies among the players. To obtain game playing skills, the players must practice repeatedly.

Figure 3 shows the screen shots in the shooter game, which we developed. In this game, player uses two different methods of attack, Gatling gun (Figure 3 (c)) and Lock-on laser (Figure 3 (a)(b)). The Gatling gun is a quick and out-range attack method. When the player uses the Gatling gun to destroy the enemies, the game score is a minimum. The lock-on laser is powerful but needs a lock-on procedure near the enemies. When the player uses a lock-on laser to destroy enemies, the game score increases exponentially with the number of lock-on targets destroyed at the same time. The player tries to obtain as high a game score as possible by selectively using the two different attack methods. Of course, as the enemies are attacking the player, the lock-on laser cannot always be used to survive in the game.

In this experiment, the participants were only trained for the first stage of the shooter game. The stage was segmented into eight parts. The patterns of combinations and the movements of enemies were different in each part. The eighth part was the battle with the final boss enemy. There was a relaxation period after each part. The average clear time was designed to be about 150 seconds. The patterns are shown in Figure 4. The first part and the fourth part are relatively easy because the density of enemies is sparse and the moving patterns are simple. Since the densities of enemies in the second and third parts are thick and the moving patterns are complex, they are relatively difficult. From the fifth to the seventh parts, the densities of enemies are relatively sparse and the moving patterns are relatively simple, but special techniques are needed to get high scores. When the player used a suitable approach in any part, his/her score was several times higher than that obtained using an inappropriate procedure. When the enemies hit the player three times, the game was over. When the boss enemy was destroyed, the game was cleared. After the game

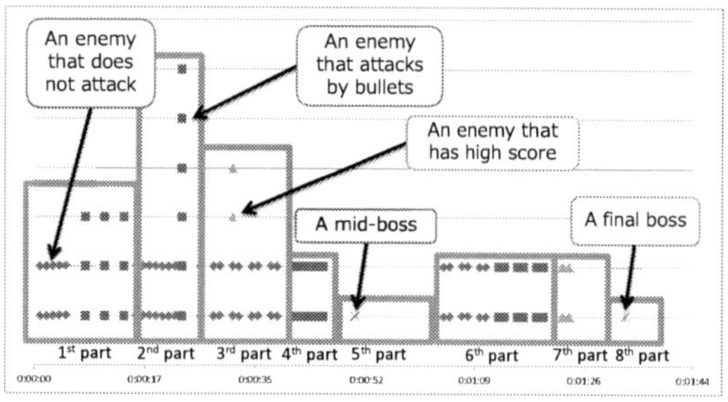

Figure 4. The outline of the enemy patterns in the shooting game.

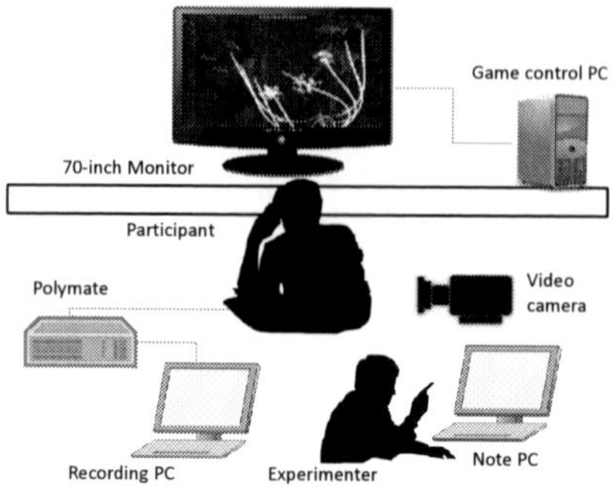

Figure 5. The experimental environment.

was over or cleared, the participants could confirm their score. They were instructed what behavior produced a high score, but it was difficult to carry out such behavior during the game. The participants acquired game playing skills through repeated practice. The game playing skill levels that we assumed are as follows.

Basic: A player can clear the game stage.
Senior: A player can get high scores in easy parts of the game stage.
Expert: A player can find the way to get high scores in difficult parts, but he/she sometimes fail.
Master: A player can get high scores in all parts of the game stage.

The easy parts are the first, fourth and seventh parts because a player can easily find the way to get high score. When a player can get high score in these parts, his/her skill is a senior level. The difficult parts are the second, third, fifth and sixth parts. The types of difficulties are different between the second and third parts and the fifth and sixth parts. In the second and third parts, a player has to deal with enemies of different behaviors in parallel. In the fifth and sixth parts, a player has to deal with enemies according to a difficult specific procedure. When a player can get high score in either type of parts, his/her skill is an expert level. When a player can get high score in all parts, his/her skill is a master level.

B. Experimental set-up

The experimental set-up is shown in Figure 5. Each participant sat in front of a 70-inch monitor that displayed the game. The participants used a joystick with two buttons for controlling the player character. A video camera was placed behind the participant to record his/her behavior and the game playing screen. The participant's voice was recorded using microphones of the video camera. Polymate was used to measure SCR and the electrocardiogram. SCR was measured with electrodes attached to the first and third fingers of the participant's non-dominant hand. The electrocardiogram was measured by connecting electrodes with paste to the participant's left side, the center of the chest, and both ears for ground and reference. The experimenter sat out of view of the participant and made notes about the participant's behavior and his/her subjective feeling of difficulty during each part. The experimenter was an expert at the game with experience teaching novice players how to get a high score and thus knew the points where to get a higher score and where is difficult for novice players.

C. Participants

The participants in the experiment were 21 male undergraduate students between the ages of 18 and 25 (with a mean age of 21.7). We eliminated the data of four participants, two of whom did not acquire the skills because their game scores did not increase throughout the experiment, and for two of whom we failed to measure the physiological indices because the electrodes detached. Therefore, the data of 17 participants were used for analysis. The participants repeatedly played the game in the experiment a total of 50 times; the total duration of the experiment was about 240 minutes. Therefore, we could obtain 850 game playing samples in all for analyses.

D. Procedure of the experiment

The experiment was divided into two sessions in the middle, and the participants took a long rest (15 minutes or more) between sessions because playing the game and acquiring the skills required heavy concentration. Each participant was briefly instructed on the experimental procedure. Electrodes for measuring SCR and the LF/HF electrocardiogram values were then attached to the participant's left hand and chest. The participant then played practice games twice. After a two-minute relaxation period, the experimenter started the video cameras and began recording the physiological indices, whereupon the participant began an experimental session. The participant played the game until it was over or cleared, then relaxed for 30 seconds. Moreover, the participants rested for 3 minutes every 10 games, during which time the experimenter scored the participant's feeling of difficulty subjectively and interviewed him/her about this sentiment at each part of the stage. In the first session of the experiment, the participants played 30 games because many games ended in the middle of the stage in the first session.

E. Analysis

We analyzed the data obtained from the experiment. We mainly focused on the data of physiological indices.

1) The relationship between participant's statuses and the physiological indices: The data obtained in the experiment are explained below.

1) The game score of each game play
2) The game score of each part of the stage
3) The feeling of difficulty as scored by the experimenter

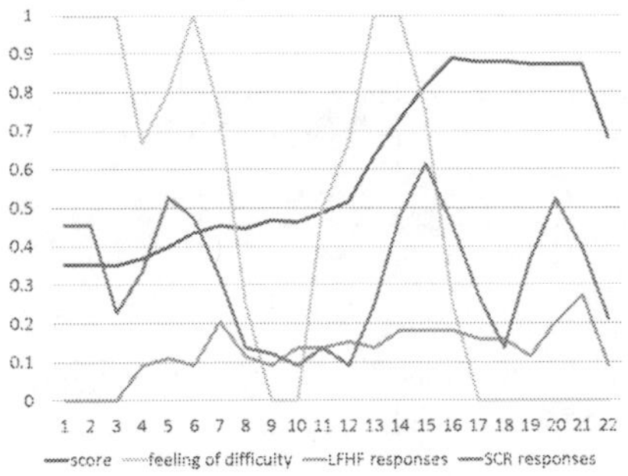

Figure 6. An example of analyzed data.

4) The physiological indices during the game play

We used 2 as the task performance and 3 as the feeling of difficulty in our model. In this analysis, we used the simple moving average (calculated from data for the previous five games and shifted by two games) for the data from 2 and 4 . The data from 3 served as categorical data to identify the feeling of difficulty. For example, the first data point was calculated from the data for games 1–5, the second data point from the data for games 3–7, and so on. Figure 6 shows an example of the analyzed data. A horizontal axis shows the data number of the simple moving average, and a vertical axis shows the ratio of the each value in each part. The task performance was assigned the value of the score for each part, divided by the maximum score for that part. The feeling of difficulty took on the values +1, i.e., felt difficulty at that part of the stage, and 0, i.e., did not feel difficulty at that part of the stage. The physiological responses were measured as the total time over the threshold (LF/HF: 3.0, SCR: 15.5) divided by the total time of that part.

We did not analyze the eighth part because we could not objectively segment the battle with the boss enemy. Thus, for this section we analyzed the first seven parts.

2) The relationship between the feeling of difficulty and the physiological indices: We calculated Spearman's rank correlation coefficients between the values of the task performance (TP) and the values of physiological responses (LF/HF and SCR), and those between the values of the feeling of difficulty (FD) and the values of physiological responses. The results are shown in Table I. In the second part, there is a weak negative correlation between the values of the task performance and the SCR responses, meaning that the scores themselves were not so strongly related to the participants' mental states. On the other hand, there are weak positive correlations between the values of the feeling of difficulty and the SCR responses in three out of the seven parts, and moderate correlations in three other parts. This confirms that there is a relationship between the feeling of difficulty and SCR.

We also performed Mann-Whitney U tests to compare the physiological responses when participants felt difficulty with the task and those when they did not. The numbers of samples are as follows (part number - number of data points where participants felt difficulty : number of other data points); the first part - 250 : 597, the second part - 370 : 472, the third part - 127 : 704, the fourth part - 575 : 275, the fifth part - 449 : 274, the sixth part - 248 : 417, and the seventh part - 216 : 395. The results are shown in Table II. In this, only the p-values under 0.1 are shown easy to see. The hyphen ("-") means that the p-value is over 0.1. There is a significant difference in the LF/HF responses only in the second part. On the other hand, there are significant differences in SCR responses in all the parts. These results indicate that SCR is related to the feeling of difficulty.

TABLE I. CORRELATION COEFFICIENTS BETWEEN THE VALUES OF THE TP AND THE PHYSIOLOGICAL RESPONSES, AND THOSE BETWEEN THE VALUES OF THE FD AND THE PHYSIOLOGICAL RESPONSES.

	LF/HF and TP	SCR and TP	LF/HF and FD	SCR and FD
1st part	-0.064	-0.189	0.042	0.354
2nd part	-0.084	-0.212	0.065	0.462
3rd part	-0.063	-0.029	-0.035	0.136
4th part	-0.018	-0.041	-0.011	0.419
5th part	0.04	0.07	0.017	0.318
6th part	-0.006	-0.065	0.036	0.5
7th part	-0.045	-0.009	-0.042	0.317

TABLE II. RESULTS OF MANN-WHITNEY U TESTS BETWEEN DATA IN THE NON-FD AND THE FD.

	LF/HF between FD and non-FD	SCR between FD and non-FD
1st part	-	0.0000
2nd part	0.0594	0.0000
3rd part	-	0.0001
4th part	-	0.0000
5th part	-	0.0000
6th part	-	0.0000
7th part	-	0.0000

In our previous study, we reported that SCR relatively reflected intrinsic stress (concentrating on the task, considering it, and so on) and that LF/HF relatively reflected the effects by extrinsic stimuli [21]. We expected that SCR would respond when the participant considered how to destroy enemies and what a bad point was in prior game plays, and therefore that we would find a relationship between the feeling of difficulty and SCR. However, we expected that the LF/HF would respond when the participant encountered impressive game events such as getting a high score or making a mistake. Impressive game events occurred independently of task performance and the feeling of difficulty. Therefore, we could not find a relationship with LF/HF.

3) The differences of the physiological indices between sections where task performance rapidly increased and others: If SCR reflects intrinsic stress, there is a link between SCR responses and changes of task performance, especially in the parts of the task where it is important to plan how to destroy enemies and get high scores. On the other hand, if LF/HF reflects the effects of external stimuli, LF/HF responses change whether the participants' technique is optimized or not. We thus expect that the values of the LF/HF responses and the SCR responses in sections where task performance rapidly increased are important for assessing the skill acquisition process in some parts of the stage. We refer to a section where task performance rapidly rose as a "Rapidly Increasing Section" (RIS), which is defined as a section in which task performance increases by more than 0.1 for a continuous two or more sections.

We compared the LF/HF responses and the SCR responses between the values in RISs and the other sections. We expected

TABLE III. Results of the Mann-Whitney U test between data of LF/HF in non-RIS and that in RIS.

	The whole part	The 1st half	The 2nd half
1st part	-	-	0.0575
2nd part	0.0020	0.0021	0.0474
3rd part	-	-	-
4th part	-	-	-
5th part	0.0835	-	-
6th part	-	-	-
7th part	-	-	0.0491

TABLE IV. Results of the Mann-Whitney U test between data of SCR in non-RIS and that in RIS.

	The whole part	The 1st half	The 2nd half
1st part	0.0004	0.0031	0.0004
2nd part	-	-	-
3rd part	0.0094	0.0012	-
4th part	-	-	-
5th part	0.0135	-	0.0000
6th part	-	-	-
7th part	0.0011	0.0147	0.0009

TABLE V. Results of the Mann-Whitney U test between data of LF/HF in non-PS and that in PS.

	The whole part	The 1st half	The 2nd half
1st part	-	0.0066	0.0905
2nd part	-	-	-
3rd part	-	-	-
4th part	0.0246	0.0576	0.0957
5th part	-	-	-
6th part	0.0498	0.0829	0.0983
7th part	0.0821	-	0.0338

TABLE VI. Results of the Mann-Whitney U test between data of SCR in non-PS and that in PS.

	The whole part	The 1st half	The 2nd half
1st part	0.0900	-	0.0822
2nd part	-	-	-
3rd part	0.0082	0.0636	0.0231
4th part	-	-	-
5th part	-	-	-
6th part	-	0.0467	0.0898
7th part	-	-	-

that the meaning of the physiological responses was different between the first and second halves of the part of the stage. The physiological responses in the first half show whether the participant had a concrete plan for scoring in advance. The physiological responses in the second half show whether the participant was used to performing the task in the part. We thus calculated the values of the LF/HF responses and the SCR responses in the whole game part and in both halves. We performed Mann-Whitney U tests for each set of values of the physiological indices. The numbers of samples are as follows (part number - number of data points in the rapid increasing section : number of other data points); the first part - 139 : 275, the second part - 128 : 263, the third part - 145 : 246, the fourth part - 94 : 297, the fifth part - 93 : 292, the sixth part - 134 : 250, and the seventh part - 101 : 271. The results are shown in Table III and Table IV. In the tables, only the p-values under 0.1 are shown easy to see. The hyphen ("-") means that the p-value is over 0.1. In the LF/HF responses for the whole game part, there is a significant difference in the second part and a marginally significant difference in the fifth part. In the SCR responses for the whole game part, there are significant differences in the first, third, fifth, and seventh parts.

Meanwhile, for the physiological responses in the first and second halves of the game part, the patterns of significant differences are different. For example, in the fifth part, there is no significant difference in the first half of the SCR responses. The task of the fifth part is to destroy a mid-boss enemy, and since the behavior of the mid boss becomes complex over time, it is reasonable that the player executes trial and error in the second half. In the case of the second part, there is a significant difference only in the LF/HF responses. In the second part, the patterns of the enemies change in the middle. Since the player has to respond the changes reactively, it is reasonable that the player takes care about extrinsic stimuli.

To sum up, the results suggest that we may distinguish the task features in the skill acquisition process using the transitions of task performance and the different patterns of the physiological responses.

4) The differences of the physiological indices between sections where task performance decreased and others: The RIS is a good example in the skill acquisition process. However, task performance sometimes decreases because the learner tries to optimize his/her behavior to the task. During optimization, the learner mainly pays attention to how to respond to the external stimuli. We thus expect that the values of the LF/HF responses and the SCR responses in such a section will fall. We refer to the section where the task performance falls into a "Plateau Section" (PS), defined as a section in which task performance stays constant or decreases continuously for two or more sections.

We compared the values of the LF/HF responses and the SCR responses between the plateau sections and the other sections. In the same way as with the RISs, we calculated the values of the LF/HF responses and the SCR responses in the whole game part and the two halves of the game part. We performed Mann-Whitney U tests on each set of values of physiological indices. The numbers of samples are as follows (part number - number of data points in the plateau section : number of other data points); the first part - 242 : 172, the second part - 218 : 173, the third part - 228 : 163, the fourth part - 256 : 135, the fifth part - 262 : 129, the sixth part - 243 : 141, and the seventh part - 249 : 123. The results are shown in the Table V and Table VI. In the tables, only the p-values under 0.1 are shown easy to see. The hyphen ("-") means that the p-value is over 0.1. Unlike the previous analysis, there is a major difference in the patterns of the significant differences in the whole and the two halves of the plateau section. In the fourth and sixth parts, there are significant differences for the whole plateau sections. On the other hand, in the first and seventh parts, there are significant differences in the first or second half of the plateau sections. In addition, we can find more differences in the LF/HF responses than in the SCR responses, suggesting that the learners may try to optimize their behavior in the short subtasks in the first and seventh parts in the plateau section. This is supporting evidence that a situation may arise where task performance does not increase even when the learner acquires sub-skills.

In this study, we know the maximum game scores in each part of the stage. Therefore, we can determine that learners try to optimize their behavior or that they are performing tasks through inertia. The plateau section is important because it is the preparation phase for the skill acquisition process. However, in terms of general skill tasks, we cannot know the maximum performance of the endeavor. Hence, we propose a method to identify plateau sections using task performance and the learner's mental state.

F. The estimation of participant's statuses using the physiological indices

As mentioned above, we can find some significant differences in the sections with the feeling of difficulty, the rapid increasing sections, and the plateau sections, which suggests that the physiological indices are clues to estimate the participant's status in skill acquisition. Of course, other clues are needed for accurate estimation, such as task structures, the basic abilities of the learners, and the details of the mental states of the learners. However, as a first step, we try to estimate the three states in the skill acquisition: whether the learner feels difficulty in a particular part of the task, whether a particular part of the task is a rapid increasing section or not, and whether a particular part of the task is a plateau section or not. To classify the states, we used Support Vector Machine (SVM) Modeling with radial basis function kernels. SVMs are supervised learning models. SVMs have been introduced for solving pattern recognition problems [22][23]. The SVM learning algorithm builds a model that assigns new cases to a certain category or another category. SVM is a non-probabilistic binary linear classifier. SVM uses a kernel function which maps the given data into a different space; the separations can be made even with very complex boundaries. The hyperplane algorithm is a way to create non-linear classifier by applying the kernel trick to maximum margin hyperplanes.

We calculated the values of the LF/HF responses and the SCR responses in the whole game part and the two halves. In addition, we also calculated the values of the mean and the standard deviation of LF/HF and SCR. To estimate the feeling of difficulty, the values were calculated for each trial of the experiment. When estimating the rapid increasing section and the plateau section, the values of the simple moving average (calculated from data for the previous five games and shifted by two games) were calculated. The 18 variables were used for SVM with radial basis function kernels. The variables were selected by the stepwise method. In this analysis, we performed leave-one-game-part-data-out cross validation for SVM classification using data from which the data of one game part were removed. The SVM model was applied to the data for each game part and the F-measure and accuracy were calculated. In the following tables, each row shows the F-measure and accuracy resulting from SVM classification using data from which the data for the game part for the row were removed. For example, the data in the second row and the fourth column of the table shows the result of applying SVM model to the data of the fourth part, calculated using the data from which the data for the second part were removed. Therefore, the diagonal values are the results of the cross validations. In the following table, the diagonal values are displayed in a bold typeface.

TABLE VII. F-MEASURES OF SVM CLASSIFICATIONS FOR THE ESTIMATION OF THE FEELING OF DIFFICULTY.

	1st	2nd	3rd	4th	5th	6th	7th
1st part	**0.523**	0.708	0.365	0.802	0.797	0.720	0.627
2nd part	0.610	**0.634**	0.358	0.810	0.802	0.717	0.638
3rd part	0.616	0.713	**0.293**	0.817	0.811	0.722	0.626
4th part	0.628	0.729	0.352	**0.741**	0.805	0.749	0.647
5th part	0.661	0.711	0.366	0.826	**0.723**	0.742	0.646
6th part	0.611	0.684	0.368	0.804	0.807	**0.656**	0.621
7th part	0.599	0.706	0.365	0.826	0.813	0.741	**0.569**

TABLE VIII. ACCURACIES OF SVM CLASSIFICATIONS FOR THE ESTIMATION OF THE FEELING OF DIFFICULTY.

	1st	2nd	3rd	4th	5th	6th	7th
1st part	**0.694**	0.757	0.740	0.749	0.736	0.789	0.691
2nd part	0.758	**0.702**	0.750	0.756	0.748	0.788	0.709
3rd part	0.750	0.752	**0.657**	0.760	0.754	0.780	0.679
4th part	0.747	0.765	0.708	**0.675**	0.734	0.800	0.700
5th part	0.780	0.751	0.724	0.770	**0.651**	0.791	0.697
6th part	0.759	0.736	0.752	0.749	0.754	**0.743**	0.694
7th part	0.744	0.752	0.732	0.772	0.757	0.798	**0.604**

1) The estimation of the feeling of difficulty: Table VII and Table VIII show the results of SVM classifications. The numbers of samples are same as in the section of "The relationship between the feeling of difficulty and the physiological indices". The diagonal values show results when the data not used to calculate the SVM model. As a result, the values of F-measure were over 0.6 in most cases and over 0.7 in 25/49 cases. The values of the accuracy were over 0.7 when the data used to calculate the SVM model were classified and over 0.65 when the data not used to calculate the SVM model (except the data in the seventh part) were classified. The results suggest the possibility of estimating the feeling of difficulty using physiological indices.

From the results, we can find that the F-measures are relatively low in the data of the third part, but the accuracies are not so low. This means that it is hard to classify the data points in the rapid increasing section. The enemy pattern in the third part was a little different from other part; two enemies that has high score parallelly appeared with another type of enemies in the part. Since, in this part, there are some strategies to get score, the stress of the participant was changed depending on the adopted strategy. In addition, depending on the adopted strategy, the feeling of difficulty of the participant was changed, but it was difficult to identify the adopted strategy. Therefore, we expect that the data of the feeling of difficulty as scored by the experimenter included some errors.

2) The estimation of the RIS: Table IX and Table X show the results of SVM classifications. The numbers of samples are same as in the section of "The differences of the physiological indices between sections where task performance rapidly increased and others". The diagonal values show results when the data not used to calculate the SVM model. As a result, the values of the F-measure and accuracy were low when the data not used to calculate the SVM model were classified. The values of the F-measure and the accuracy were over 0.5 and over 0.65, respectively, when the data used to calculate the SVM model were classified. The results suggest that the generalization capability was low, but that there is a

TABLE IX. F-MEASURES OF SVM CLASSIFICATIONS FOR THE ESTIMATION OF THE RIS.

	1st	2nd	3rd	4th	5th	6th	7th
1st part	**0.417**	0.573	0.553	0.535	0.638	0.634	0.655
2nd part	0.606	**0.401**	0.616	0.523	0.600	0.618	0.625
3rd part	0.584	0.556	**0.406**	0.535	0.651	0.634	0.639
4th part	0.596	0.587	0.592	**0.283**	0.642	0.618	0.675
5th part	0.615	0.578	0.614	0.568	**0.314**	0.644	0.658
6th part	0.601	0.586	0.601	0.531	0.624	**0.305**	0.702
7th part	0.589	0.599	0.563	0.576	0.591	0.658	**0.331**

TABLE X. ACCURACIES OF SVM CLASSIFICATIONS FOR THE ESTIMATION OF THE RIS.

	1st	2nd	3rd	4th	5th	6th	7th
1st part	**0.534**	0.665	0.632	0.693	0.760	0.708	0.782
2nd part	0.674	**0.435**	0.627	0.655	0.724	0.672	0.728
3rd part	0.684	0.652	**0.565**	0.693	0.770	0.711	0.766
4th part	0.696	0.680	0.683	**0.430**	0.754	0.716	0.790
5th part	0.686	0.634	0.662	0.731	**0.476**	0.729	0.780
6th part	0.676	0.650	0.650	0.670	0.731	**0.503**	0.806
7th part	0.667	0.655	0.642	0.706	0.734	0.729	**0.522**

TABLE XI. F-MEASURES OF SVM CLASSIFICATIONS FOR THE ESTIMATION OF THE PS.

	1st	2nd	3rd	4th	5th	6th	7th
1st part	**0.462**	0.584	0.659	0.691	0.738	0.696	0.803
2nd part	0.684	**0.538**	0.656	0.676	0.733	0.720	0.746
3rd part	0.647	0.541	**0.495**	0.712	0.759	0.682	0.736
4th part	0.675	0.637	0.701	**0.563**	0.727	0.754	0.761
5th part	0.683	0.656	0.679	0.745	**0.650**	0.723	0.809
6th part	0.664	0.589	0.606	0.664	0.734	**0.636**	0.777
7th part	0.695	0.602	0.618	0.748	0.765	0.747	**0.620**

TABLE XII. ACCURACIES OF SVM CLASSIFICATIONS FOR THE ESTIMATION OF THE PS.

	1st	2nd	3rd	4th	5th	6th	7th
1st part	**0.488**	0.606	0.642	0.657	0.675	0.641	0.758
2nd part	0.640	**0.504**	0.619	0.634	0.662	0.661	0.688
3rd part	0.626	0.558	**0.504**	0.668	0.675	0.646	0.680
4th part	0.616	0.601	0.645	**0.504**	0.668	0.688	0.702
5th part	0.633	0.609	0.616	0.688	**0.545**	0.656	0.753
6th part	0.597	0.542	0.560	0.611	0.650	**0.549**	0.715
7th part	0.640	0.593	0.575	0.680	0.675	0.682	**0.530**

possibility of estimating the rapid increasing section using the data used to calculate the SVM model. Of course, the values of accuracy when the data used to calculate the SVM model were classified were not sufficient in some parts. We have to consider more details in the participant's mental states and the conditions of the task.

3) The estimation of the PS: Table XI and Table XII show the results of SVM classifications. The numbers of samples are same as in the section of "The differences of the physiological indices between sections where task performance decreased and others". The diagonal values show results when the data not used to calculate the SVM model. As a result, the values of the F-measure and accuracy were not so high when the data not used to calculate the SVM model were classified. The values of F-measure and accuracy in the plateau section were lower than those in the rapid increasing section. One reason is that situations in which learners try to optimize their behavior and situations in which they are performing tasks through inertia are mixed. More detailed teacher data are needed to improve the classification.

V. DISCUSSION

This study aimed to develop a method to estimate the learner's skill level based on task performance and his/her mental states. To achieve this, we conducted an experiment to obtain data from the subjective reports of a feeling of difficulty and the physiological indices during the skill acquisition process. We then confirmed the relationship between the subjective reports of the feeling of difficulty and the physiological indices. In addition, we suggested an approach to identify the statuses of the skill acquisition process, e.g., the learner feels difficulty and the task performance is rapidly increasing, via the measured physiological indices. Concretely, we classified the trials with a feeling of difficulty in the experiment using SVM. As a result, the values of accuracy were over 0.7 when the data used to calculate the SVM model were classified and over 0.65 when the rest of the data not used to calculate the SVM model were classified. When we classified the trials in the rapid increasing section, the values of accuracy were over 0.65 when the data used to calculate the SVM model were classified. However, when we classified the trials in the plateau section, the values of accuracy were low. Although we have to refer to the teacher data for more accurate classification, we suggest that there is a possibility of estimating the statuses of the skill acquisition process.

In the results of the Mann-Whitney U tests, there is a different tendency between the results for the feeling of difficulty and the results of the rapid increasing and plateau sections. In the results of the feeling of difficulty, the SCR response is significantly different in all parts of the game but the LF/HF response is not. On the other hand, in the results of the rapid increasing section and the plateau section, the SCR response is not significantly different in certain parts of the game, while the LF/HF response is significantly different in such parts. During the skill acquisition process, learners execute actions by trial and error, receive feedback, and improve their behavior based on subjective and objective evaluations. The learners feel difficulty when they evaluate their behavior after finishing the task or sub-task and thus are not influenced by the extrinsic stimuli in the task, i.e., their LF/HF values are not changed. When the learners improve their behavior in real time, they have to pay attention to the extrinsic stimuli in the task depending on the statuses of the task, e.g., the patterns of the enemies and whether they already have a plan to score. The patterns of the significant differences in the SCR and LF/HF responses reflect the statuses of the task.

A comparison of the results of the Mann-Whitney U tests and those of the SVM classification shows that the physiological responses showing significant differences in the Mann-Whitney U tests do not directly contribute to the SVM classification. One reason is that the distributions of the physiological responses overlap between the statuses. For example, the results for the third part in the feeling of difficulty show no difference from the others, but it is difficult to classify the data for the third part using the SVM. Conversely, it is not so difficult to classify the data for the sixth part in the rapid increasing section using the SVM, but there is no significant difference in the physiological responses. These suggest that the skill acquisition of the skill task has complex processes related to different factors and only one particular parameter does not suffice to show the status of the skill acquisition. Meanwhile, we suggest that the multiple physiological indices are important clues to the status of the skill acquisition process.

The most important contribution of this study is that we experimentally analyzed the effects of the learner's mental states based on the physiological indices. In a traditional skill acquisition model, the impacts of the learner's mental states are conceptually proposed but not confirmed objectively. Of course, our study has some limitations. The most serious limitation was that we could not achieve sufficient F-measure and accuracy in the SVM classification when the data not used to calculate the SVM model were classified. One reason is that the important points for acquiring the skill are different for each part of the game. For example, in the third part, participants need to pay attention to the different types of enemies in parallel. For accurate classification, it is an important piece of information, which enemy was paid attention to, but such a clue could not be included in the SVM classification. Another limitation was the segmentation of the task. In this research, we segmented the game into eight parts (including the final boss part). However, the segmentations were not sufficient for accurate classification, which is a task awaiting future work on realizing a skill acquisition support system.

VI. Conclusion and Future Work

The purpose of this study was to experimentally investigate whether the statuses of the skill acquisition process could be estimated by the physiological indices used to estimate the learner's mental states. For this purpose, we conducted an experiment to obtain the data of the physiological indices and a subjective report of the feeling of difficulty during the skill acquisition task. As a result of the analysis, we confirmed the relationship between the participant's statuses (feeling of difficulty, the rapid increasing section, and the plateau section) and the physiological indices. In addition, we classified the trials having a feeling of difficulty in the experiment using SVM. Overall, we could show that the physiological indices are helpful clues to estimate the status of the skill acquisition process. In future research, we will obtain more teacher data for machine learning techniques and analyze the different features of the skill acquisition task. In addition, we will extend the framework of experiments to different tasks that include the different features of the skill acquisition task.

VII. Acknowledgement

This research is supported by Grant-in-Aid for Young Scientists (B) (KAKENHI No. 16K21113), and Grant-in-Aid for Scientific Research on Innovative Areas (KAKENHI No. 26118002) from the Ministry of Education, Culture, Sports, Science and Technology of Japan.

References

[1] Y. Ohmoto, T. Matsuda, and T. Nishida, "Experimentally analyzing the skill acquisition model using task performance and physiological indices," in COGNITIVE 2017: The ninth International Conference on Advanced Cognitive Technologies and Applications. IARIA, 2017, pp. 44–49.

[2] S. E. Dreyfus and H. L. Dreyfus, "A five-stage model of the mental activities involved in directed skill acquisition," DTIC Document, Tech. Rep., 1980.

[3] J. Tranquillo and M. Stecker, "Using intrinsic and extrinsic motivation in continuing professional education," Surgical neurology international, vol. 7, no. Suppl 7, 2016, p. S197.

[4] K. Kraiger, J. K. Ford, and E. Salas, "Application of cognitive, skill-based, and affective theories of learning outcomes to new methods of training evaluation," Journal of applied psychology, vol. 78, no. 2, 1993, p. 311.

[5] T. R. Mitchell, H. Hopper, D. Daniels, J. George-Falvy, and L. R. James, "Predicting self-efficacy and performance during skill acquisition," Journal of Applied Psychology, vol. 79, no. 4, 1994, p. 506.

[6] F. Gobet and P. Chassy, "Expertise and intuition: A tale of three theories," Minds and Machines, vol. 19, no. 2, 2009, pp. 151–180.

[7] E. A. Day, W. Arthur Jr, and D. Gettman, "Knowledge structures and the acquisition of a complex skill," Journal of applied psychology, vol. 86, no. 5, 2001, p. 1022.

[8] J. Langan-Fox, K. Armstrong, N. Balvin, and J. Anglim, "Process in skill acquisition: Motivation, interruptions, memory, affective states, and metacognition," Australian Psychologist, vol. 37, no. 2, 2002, pp. 104–117.

[9] P. M. Fitts and M. I. Posner, "Human performance," 1967.

[10] J. R. Anderson, "Acquisition of cognitive skill," Psychological review, vol. 89, no. 4, 1982, p. 369.

[11] R. M. Shiffrin and W. Schneider, "Controlled and automatic human information processing: Ii. perceptual learning, automatic attending and a general theory," Psychological review, vol. 84, no. 2, 1977, p. 127.

[12] C. Konrad, G. Schupfer, M. Wietlisbach, and H. Gerber, "Learning manual skills in anesthesiology: is there a recommended number of cases for anesthetic procedures?" Anesthesia & Analgesia, vol. 86, no. 3, 1998, pp. 635–639.

[13] F. E. Ritter and L. J. Schooler, "The learning curve," International encyclopedia of the social and behavioral sciences, vol. 13, 2001, pp. 8602–8605.

[14] T. P. Grantcharov and P. Funch-Jensen, "Can everyone achieve proficiency with the laparoscopic technique? learning curve patterns in technical skills acquisition," The American Journal of Surgery, vol. 197, no. 4, 2009, pp. 447–449.

[15] J. T. Patterson, A. Hart, S. Hansen, M. J. Carter, and D. Ditor, "Measuring investment in learning: Can electrocardiogram provide an indication of cognitive effort during learning?" Perceptual and motor skills, vol. 122, no. 2, 2016, pp. 375–394.

[16] A. D. Walker, E. R. Muth, F. S. Switzer III, and P. J. Rosopa, "Predicting team performance in a dynamic environment: A team psychophysiological approach to measuring cognitive readiness," Journal of Cognitive Engineering and Decision Making, vol. 7, no. 1, 2013, pp. 69–82.

[17] P. Parsinejad, "Inferring mental workload changes of subjects unfamiliar with a touch screen game through physiological and behavioral measurements," Ph.D. dissertation, Northeastern University, 2016.

[18] E. F. Bartholomew, F. Martini, and W. B. Ober, Essentials of anatomy & physiology. Benjamin Cummings, 2007.

[19] K. Hugdahl, Psychophysiology: The mind-body perspective. Harvard University Press, 1995.

[20] B. C. Lacey and J. I. Lacey, "Two-way communication between the heart and the brain: Significance of time within the cardiac cycle," American Psychologist, vol. 33, no. 2, 1978, p. 99.

[21] Y. Ohmoto, S. Takeda, and T. Nishida, "Distinction of intrinsic and extrinsic stress in an exercise game by combining multiple physiological indices," in Proceedings of the seventh International Conference on Games and Virtual Worlds for Serious Applications (VS-Games). IEEE, 2015, pp. 1–4.

[22] V. Vapnik, "The support vector method of function estimation," Nonlinear modeling: Advanced black-box techniques, vol. 55, 1998, p. 86.

[23] N. Cristianini and J. Shawe-Taylor, An introduction to support vector machines and other kernel-based learning methods. Cambridge University Press, 2000.

The Application of Validating and Dialectical Coaching Strategies in a Personalised Virtual Coach for Obese Emotional Eaters

Rationale for a Personalised Coaching System

Aranka Dol[1], Christina Bode[2], Hugo Velthuijsen[1], Lisette van Gemert-Pijnen[2], Tatjana van Strien[3,4]
[1]Institute for Communication, Media & IT, Hanzehogeschool UAS, Groningen
[2]Department of Psychology, Health and Technology, University of Twente, Enschede
[3]Behavioural Science Institute, Radboud University, Nijmegen
[4]Department of Health Sciences and the EMGO Institute for Health and Care Research, VU University, Amsterdam
e-mail: [1]{a.dol, h.velthuijsen}@pl.hanze.nl, [2]{c.bode, j.vangemert-pijnen}@utwente.nl, [3]t.vanstrien@psych.ru.nl

Abstract—**There is a growing number of eHealth interventions aiming at enhancing lifestyle to address obesity. However, the existing interventions do not take the emotional aspects of obesity into account. Forty percent of the overweight population is an emotional eater. Emotional eaters gain weight because of poor emotion regulation, not just due to bad eating habits. We aim at developing a personalised virtual coach 'Denk je zèlf!' providing support for self-regulation of emotions for obese emotional eaters. This paper presents a research study protocol on validating persuasive coaching strategies in emotion regulation, based on Dialectical Behaviour Therapy, ultimately targeting behaviour change. Our goal is to design a personalised eCoaching framework, allowing us to optimally translate successful behaviour change mechanisms and techniques, such as dialectical strategies, into personalised persuasive coaching strategies.**

Keywords-obesity; emotional eaters (emotional eating behaviour); Dialectical Behaviour Therapy; validation strategies; dialectical strategies; eHealth; mHealth; persuasive technology; virtual coach; personalised coaching.

I. Introduction

Obesity has become a major societal problem worldwide [1]-[3]. The main reason for severe overweight is an excessive intake of energy, in relation with the individual needs of a human body. Obesity is associated with poor eating habits and/or a sedentary lifestyle. A significant part of the obese population (40%) overeat due to negative emotions [4]. They suffer from food cravings and give in to food binges in response to high negative feelings or stress. There is little attention to the personal needs of emotional eaters when it comes to existing health interventions [5]-[7]. Current face-to-face treatments have insufficient recognition to affect regulation.

There is a need for self-management support and personalised coaching for emotional eaters to train themselves in recognising and self-regulating their emotions [8][9].

This paper is structured as follows: In Section II, we discuss the theoretical underpinnings of dialectical behaviour therapy and persuasive coaching strategies. Next, in Section III, we elaborate on the design rationale of a personalised virtual coach. And in Section IV we present the design of the research protocol. We finish this paper in Section V with a discussion on the expected contributions of this research study, conclusion, and future work.

II. Background

Virtual coaching systems form a broad and vivid research area [10]-[13]. Since chatbot Eliza was invented by Joseph Weizenbaum [14], many new studies have emerged. Over the last years, virtual coaches have been developed for behaviour change support, healthy lifestyle, and physical activity support [15]-[19]. Existing virtual coach applications lack systematic evaluation of coaching strategies and usually function as (tele)monitoring systems, limited to giving general feedback to the user on achieved goals and accomplished (online) assignments [20].

This paper presents the research protocol and design rationale for a personalised coaching system and the evaluation of persuasive coaching strategies for emotional eaters based on the Dialectical Behavior Therapy and Persuasive Technology.

This research study is a continuation of a previous study on the design and development of *emotion-enriched* personas [9]. The objective of developing a personalised virtual coach for obese emotional eaters is to raise their awareness about emotions, to enhance a positive change of attitude towards accepting the negative emotions they experience, and to accomplish a balance in body weight.

A. Dialectical Behavior Therapy

Dialectical Behavior Therapy (DBT) focuses on getting more control over one's own emotions by reinforcing skills in mindfulness, emotion regulation, and stress tolerance [8]. Emotion regulation is about recognising and acknowledging emotions and accepting the fact that they come and go. The behaviour change strategies within DBT are based on validation and dialectical strategies [21]. Validation is about acknowledging one's experiences or feelings, without judgment. Validation strategies suggest responding in an empathic way, by hearing another person's point of view and accepting them (and their emotions) without judging.

Dialectical strategies help to find truth (synthesis) in opposing viewpoints and encourage a change of the users' attitude and behaviour. Opposites create incongruence between belief and behaviour since the stimuli or the given information contradict each other. Dialectical strategies focus on confronting the user with a practical focus on changing problem behaviour; the key is in finding a balance between acceptance of intense feelings and emotions and the need for change by adapting feelings and emotions using emotion regulation and interpersonal effectiveness.

Emotional eaters deal with an invalidating daily life environment. With a well-balanced mix of being validated in their perception of negative emotions and a confrontation (dialectical) focussing on changing problem behaviour, DBT offers powerful mechanisms of change [23]. Our eCoaching model translates these mechanisms into persuasive features and compelling dialogues.

B. Persuasive Technology

Persuasive Technology (PT) is a significant predictor of adherence and offers strategies to reinforce the validation – and dialectical strategies' influence on attitudes and behaviours [23]. We believe that PT can assist emotional eaters in attitude change and acceptance of their own emotions. It is essential to identify the intended outcome or change of the intervention, before determining the design principles and coaching strategies that would positively contribute to the persuasiveness in any way. We selected the Persuasive Systems Design (PSD) model [24] for translating behaviour change techniques from DBT into persuasive coaching strategies [25] since PSD focuses on persuasion context and easily implementable compelling design features. We assume that the integration of the dialectical behaviour change strategies and persuasive functions (PSD model) will enhance the personalisation of the virtual coach for emotional eaters and improve the effectiveness of the intervention.

III. Design Rationale for a Personalised Virtual Coach

The virtual coach 'Develop a wise mind and counsel yourself!' (a translation of the Dutch title: 'Denk je zèlf!') in Figure 1 is an interactive and self-learning persuasive system that coaches attitude and/or long-term behaviour change by providing real-time personalised support [9] and by supplying personal feedback based on persuasive coaching strategies, originated from validation and dialectical strategies of DBT and persuasive features of the Persuasive System Design model [24]. It provides the user with practical exercises to enhance skills in mindfulness, emotion regulation and stress tolerance.

A. Components Virtual Coach

The virtual coach application consists of three different components:
1. the personalised Virtual Coach,
2. the emo.analyser,
3. a set of skills practices in mindfulness, emotion regulation and stress tolerance.

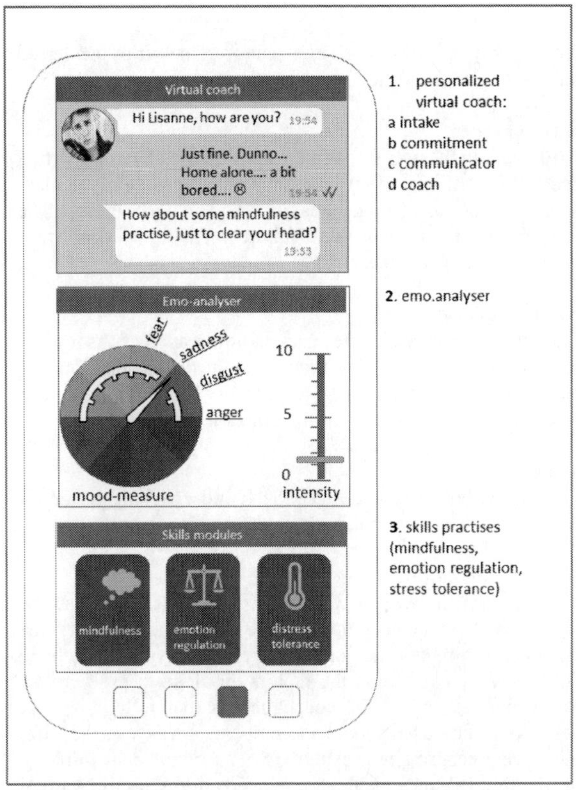

Fig. 1. Schematic overview Denk je Zelf! application

1. Personalised Virtual Coach

The virtual coach is the main part of the application. The virtual coach holds a central position in the application. It acts as a mentor or tutor throughout the application by four functions:
a. intake; b. commitment; c. communicator; and d. coach. It is always on hand for questions. After downloading and installing the application on the smartphone, the virtual coach will be nested in the Contacts list and will become a friend eventually. It communicates as the user prefers, via the private chat timeline delivered with the application, or in WhatsApp, Facebook Messenger, Slack or Telegram.

a. intake
New users have to download the application from the App Store. After installation of the application, the user is guided by the virtual coach to go to the intake module. It invites the user to register and to set a personal profile (the intake module) with information about demographics data, such as (nick) name, age, gender, weight, length, place of residence. The virtual coach wants to get to know the user to be able to provide the utmost service and personalised feedback. She will be asked to fill in questionnaires about her eating behaviour: "Are you an emotional eater?" [26] and about her personality traits: "What kind of person are you?" [27]. The virtual coach also offers features to customise the application

with favourite colour schemes, and to select a preferred picture to personify the virtual coach.

b. commitment
After leading the user through the intake procedure, the virtual coach helps the user to set personal goals and solemnly declare to quit emotional eating behaviour and to learn how to be prepared for the possibility of failure when things don't turn out to be useful at once.

c. communicator
The virtual coach welcomes users at login and acts as a point of contact. It supplies information about the application and redirects users to the other components. It provides the user feedback on the outcomes of the emo.analyser.

d. coach
The virtual coach provides feedback and coaching based on the information users put down in the emo.analyser.

2. Emo.analyser

The emo.analyser is a built-in tool to help the user to register what happened today that made them experience cravings or binges. The goal of the emo.analyser is to create awareness by examining the factors that lead to problematic eating behaviour and the consequences that follow on that behaviour. The user gets to the emo.analyser by (a). on her initiative, because she is experiencing cravings or is thinking of binging; (b). as a result of a conversation with the Virtual Coach, triggered by emotion - or eating behaviour related words, or (c). the user logged in after a reminder to visit the app and the answer to one question about her current emotional status. In a progressive scheme the user will be questioned about her eating behaviour over the past 24 hours. See Figure 2. The first step of six is to point out the problem behaviour (1). Problem behaviour varies from mindless eating to binge eating. The next step is to determine the prompting event (2) that triggered the problem behaviour. Prompting events emerge in the user's surroundings. The user specifies what made her vulnerable (3) to perform that specific behaviour. This event could have an internal or external reason, such as physical illness or being home all alone, or being invited to a birthday party. The problem behaviour and the prompting event are linked (4) by a series of thoughts, actions, bodily sensations and cognition. Once these are all identified, users are to describe the consequences (5) of their problem behaviour. This way of behaving varies from staying away from a birthday party, to feeling numb, or being depressed. The last step in the chain is to figure out how to repair (6) and reduce vulnerability for example by getting more sleep, or by not buying snack food.

The emo.analyser collects the above-mentioned data and provides feedback, shown in a graphic presentation, on the discovered behaviour patterns. By doing the exercise over and over, the user gains a better understanding of patterns in how she responds to emotional occurrences, the impact on her eating behaviour, and the consequences for her emotional well-being. Based on the context (e.g., the weather, log data on location, calendar) and data about the personality and emotional state of the user, the virtual coach application enables tailoring of real-time feedback to the individual user (A). The Virtual Coach provides the user with validating and

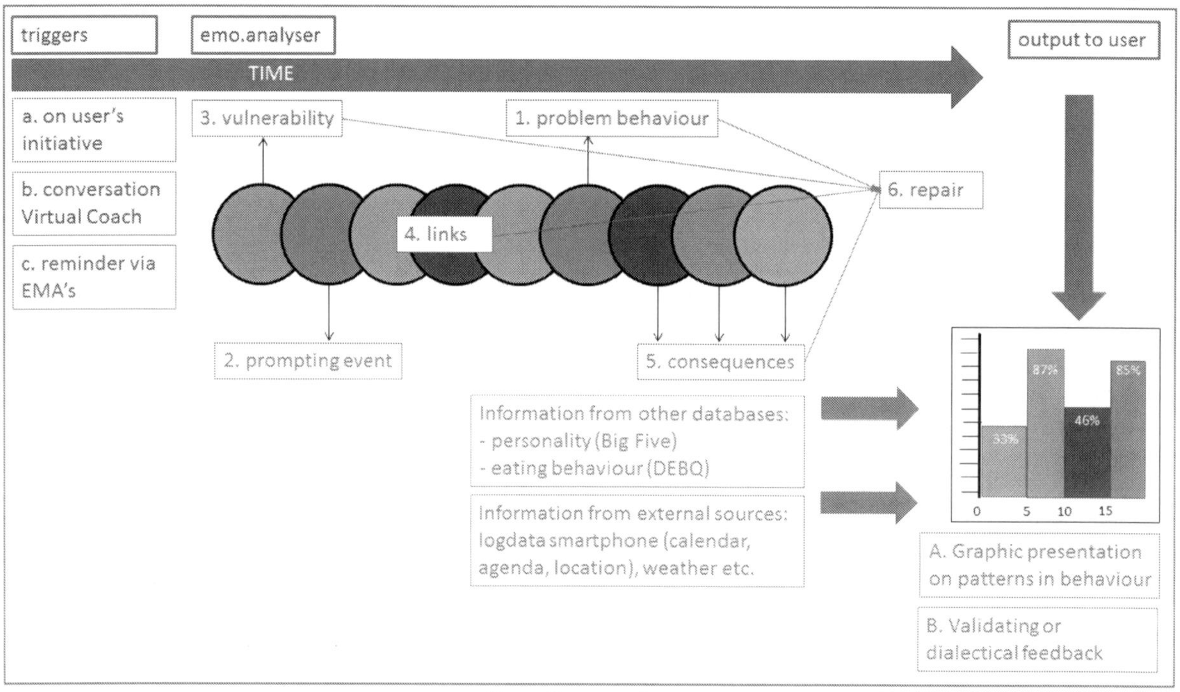

Figure 2. emo.analyser

dialectical coaching strategies (B). These strategies ratify the emotions that the user experiences, but could also put them on edge by providing statements that oppose each other. By expressing opposite sides that both hold true, a sense of imbalance might be achieved which brings about a renewed attitude towards her behaviour eventually [28]. It helps the user to monitor her emotional and behavioural patterns. Personalised coaching is articulated by providing validating or dialectical comments based on the patterns.

Validation strategies focus on 'empathic' approach suggesting to genuinely hear another person's point of view and accept them (and their emotions) just the way they are, without judging. For example, in a dialogue a (virtual) coach is clarifying the message of the user by '*reflection*': "So what I hear you saying is..." or '*direct*': "To me it makes sense that...."; or '*empathising*' the emotion "I can see you are sad"; or a cognition "Is that true, or you just think it is?"; or in a more '*cheerleading*' way "You are doing well. I am proud of you!'', and radically accept the messenger.

Dialectical means finding the synthesis out of opposites (thesis + antithesis). It is finding a balance between acceptance and change: accepting the occurrence of strong feelings and emotions making use of mindfulness and stress tolerance and adapting feelings and emotions using emotion regulation and interpersonal effectiveness. Dialectical strategies focus on confronting the user with a practical focus on changing problem behaviour. Dialectical strategies consist of several sub-strategies, such as *'entering the paradox strategy'* with "and" versus "but" responses; the use of '*natural change*' with the "That was then, this is now…" response; the *'making lemonade out of lemons strategy'* with the "That's great!" response; the *'devil's advocate technique'*, with the "Are you sure?" response, or *'extending'* with the "Wow, that's pretty serious!" response.

To make users completing the emo.analyser questionnaire on a daily basis, the virtual coach sends out a (WhatsApp or SMS) message to the user with the question whether she experienced problems with her eating behaviour, such as cravings or binges, over the past 24 hours. When answering "No" it goes to "exit" for this day.

3. Skills practises

The application contains three modules on skills practices in *mindfulness*, *emotion regulation,* and *distress tolerance*. The module mindfulness provides exercises in observing, describing and participating in reality, enhancing the quality of awareness in a person's life. The module emotion regulation contains exercises in building self-efficacy and learning to cope with difficult situations. Distress tolerance teaches skills how to survive crises and to accept reality as it is.

B. Functionalities Virtual Coach

The application Denk je zelf! contains four types of tasks: 1. Conversational features; 2. Functional features; and 3. Ecological Momentary Assessments.

1. Conversational

The conversational tasks are carried out by the Virtual Coach. The Virtual Coach (VC) is a text-based both conversational and functional in-app chatbot [29][30]. It has conversations with the user over the built-in timeline, or via WhatsApp or Facebook Messenger. As a contextual help system, it understands text from users to a certain extent, based on natural language processing and recognising wording patterns from an emotions-related lexicon. The VC is always available. It knows the user's name. It produces appropriate textual feedback and coaching according to the information that users have filled out in the emo.analyser.

2. Functional

The Virtual Coach provides users with useful information about emotional eating behaviour. It redirects users to valuable skills practices. It sends out reminders/invitations to use the emo.analyser. The user controls the whole process of *onboarding*. It starts with some basic options. The user has the ability to opt-out any time. It guides the user through the onboarding process (Fig. 3) – the whole course from first log in (1), via setting the personal profile (2), to arrival at the actual start-up screen (3), in which the user dictates how much time she spends on the intake procedure at that very moment. The more steps involved, the more knowledge about the user is collected.

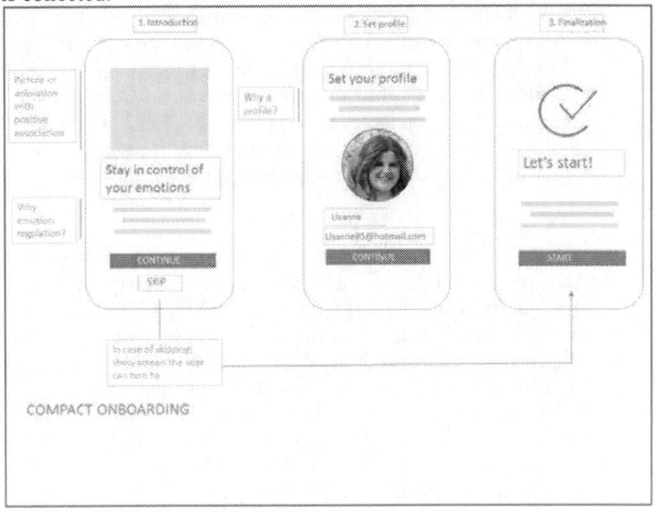

Figure 3. Compact onboarding

The Virtual Coach provides the user with visualisations of the data generated from the questionnaires at intake and from data collected by the emo.analyser. Data is stored in a local folder on the smartphone.

The option-based user interface (Fig. 4) presents the user with an overview of options with the help of which the user can respond. A multi-functional keyboard makes it easier for the user to find essential issues such as favourite skills practices, pages last visited, and fun stuff just by pushing the right button instead of typing words.

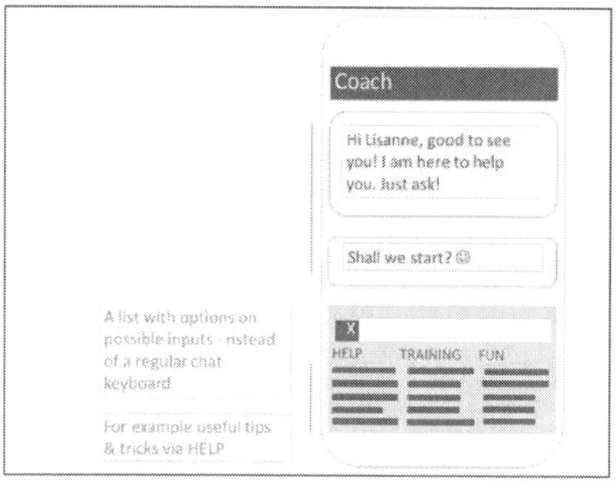

Figure 4. Option-based keyboard

3. Ecological Momentary Assessments (EMA)

Ecological Momentary Assessments are applied to systematically collect data about daily experience and feelings of the users, as well as the context of use, for instance, through a user-friendly smartphone application. Experience sampling is widely used in daily psychiatry practice [30][31], allowing to measure and record encountered feelings and emotions through mHealth applications in real time.

The Virtual Coach sends out a daily message via the built-in timeline, or via WhatsApp or Facebook Messenger inviting the user to log in into the application to answer just one question: "Did you have cravings or binges today, in the last 24 hours?". As soon as she answers that question with a "yes", she will be redirected to the emo.analyser.

IV. APPROACH AND METHODOLOGY

Before starting developing a prototype of the above-mentioned modules and functions, more research is needed to find out how the the target group experiences the validating and dialectical coaching strategies. That is why we put up an experimental study design first. For this vignette study - described under V. *A. Vignettes study* - we designed a low-fidelity concept of the virtual coach to evaluate validating and dialectical strategies with target users. This concept version lacks the functionality of the emo.analyser and the conversational functionalities. Next step will be building a working prototype of the emo.analyser. The functional criteria, as drawn up along the lines of the *Behaviour Chain Analysis* [8, pp.57-58,75] from the Dialectical Behaviour Therapy will then be coupled with the outcomes of the vignette study, being the ratified coaching strategies.

A. Vignettes study

In the design of this experimental study, participants are presented with scenarios describing the daily life of a persona [8]. A persona is a fictional representation of the target group of emotional eaters. The persona experiences the very same negative emotions and also encounters food cravings that can lead to binge eating. The realistic scenarios are based on validated personas [9]. The depicted circumstances in the vignettes [33]-[35] will appeal to the emotions and feelings of the participants in the study. See Figure 5. Three different coaching strategies are presented at random to the participants: 1. validating response, 2. confronting or dialectical response, or 3. a combination of a validating and a dialectical response.

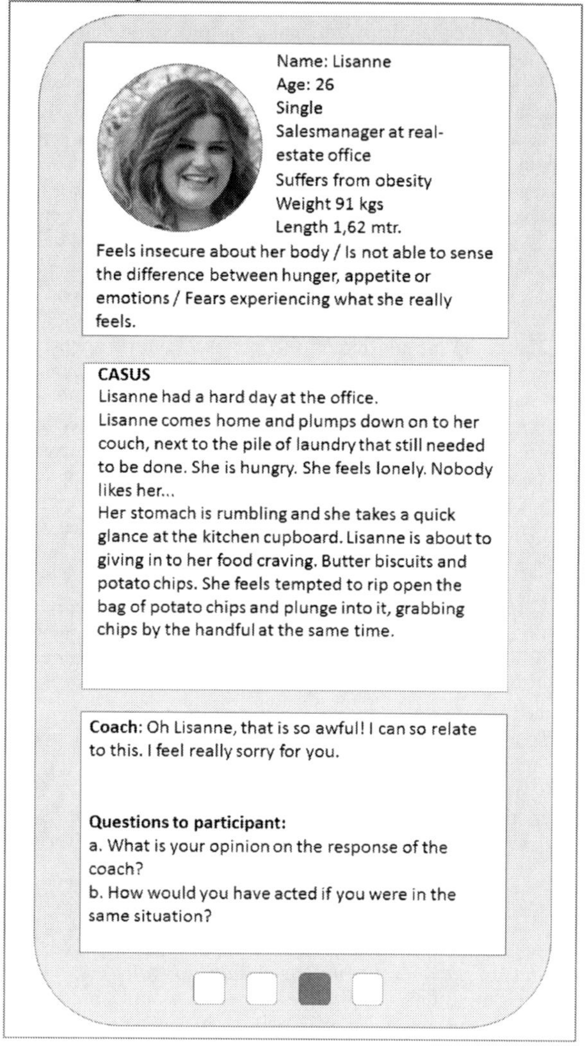

Figure 5. Vignette, example.

B. Research questions

This research study aims at answering the following research questions:

- Which coaching strategies do emotional eaters prefer just at the moment they are experiencing food cravings and are about to give in to binging?
- Is there a relationship between the participant's preferences and her personality traits?
- Is there a relationship between the emotional eating behaviour (e.g., intensity, fierceness, what negative

emotion is dominating, time of the day etc.) and the participant's preferences for a coaching strategy?

We will ask the participants to state their opinion on the given scenario. Does it seem realistic to them? Do they relate to the personas? Do they find the given feedback useful? How would they respond to the provided feedback themselves? What kind of coaching style would they prefer?

For the vignette study, we will use Qualtrics - a tool to interrogate experiences and to distribute questionnaires [36].

C. Questionnaires and measures

After answering the questions on the vignettes, participants will be referred to sequential inquiries to get to know them better: "What type of eater are you?" on eating behaviours, "Who are you?" on personality traits and "How do you feel at the moment?" on current emotions:

1. the Dutch Eating Behavior Questionnaire (DEBQ) [26] with 33 questions to determine which eating behaviour out of emotional, external, and restraint eating behaviour is prevalent, using a 5-point Likert scale (1=never; 5=very often),
2. The Quick Big Five Personality Test [27], based on the five-factor model that describes five broad trait dimensions, to identify the personality traits neuroticism, extraversion, openness to experience, agreeableness, and conscientiousness, using a 5-point Likert scale (1=does not apply; 5=applicable). The results of this questionnaire will give us more clues about what personality traits will be suitable for the application.
3. The Positive and Negative Affect Scales (PANAS) [37], containing 20 feelings and emotions the participants experience at this particular moment, using a 5-point Likert scale (1=very slightly or not at all; 5=extremely).

During the whole time span of the research study, namely seven days, some Ecological Momentary Assessments [29][30] will be carried out. Participants receive text messages (SMS) on a daily basis, with an invitation to log in to the application and answer to just one question: "Did you have cravings / have a binge today?' In case of a "YES", participants will be redirected to a questionnaire named emo.analyser, the trying-plane for the future emo.analyser application. In this questionnaire, participants will be asked to fill in what happened right before they gave in to a binge, and which emotions/feelings they were experiencing.

In this chain of events and emotions, they analyse the vulnerabilities that made them open to experiencing specific events, developing strong feelings and emotions over it, and making them giving in to binges. In case of a "NO" participants will be invited to answer some additional questions.

D. Methods

The target group consists of emotional eaters (N=45), women, age 18 years and upwards. Participants are recruited via a Dutch franchise organisation of dietitian nutritionists, who are specialised in treating emotional eating behaviours.

First, a pilot study will be conducted with 2-3 participants to validate the study protocol and to test the questionnaires. Participants are ensured in the anonymity of their participation. Data collected during this research study will be analysed for research purposes only. The data will be stored anonymously and safely on a local server, using an encrypted secure network connection (https) only the main researchers have access to.

Recruited participants are invited by email to visit the stated link. They will be asked to fill out the demographics data, such as (nick) name, age, gender, weight, length, place of residence.

Participants will be invited to participate in a semi-structured post-interview to discuss their experiences with the presented dialogues and their motivation for picking specific answers. Post-interviews will be audio recorded with permission of participants.

V. DISCUSSION, CONCLUSION AND FUTURE WORK

This paper presents the research study design to evaluate and validate coaching strategies that emotional eaters prefer in a personalised virtual coach, based on their emotional state and context. The potential contribution of this research is divided into two objectives.

First, a multidisciplinary holistic user-centered design approach [38] provides valuable insights into preferred persuasive coaching strategies of the target users for developing a meaningful and persuasive virtual coaching system for self-managing emotions. One of the primary goals of the personalised virtual coach for emotional eaters is to raise their awareness of their own emotions and to enhance a positive change of attitude towards accepting the negative emotions they experience daily. Ultimately, improving emotion regulation is expected to lead to better weight management of emotional eaters.

Second, it will gain more insights into applying the behaviour change techniques from health psychology and persuasive technology for virtual coaching application. Next step is to focus on developing virtual coaching systems that coach the user by analysing real-time behaviour by providing real-time personalised timely feedback based on user profile, events, and context. Coaching-based applications can provide self-management support for patients at times when a therapist is not available, or in the case that the user needs personalised anonymous help that is always within reach, which is the case for emotional eaters who often experience feelings of shame.

After conducting the vignette study we will be able to provide personalised and customised feedback and coaching based on experienced and digested emotions and feelings that users have entrusted to the emo.analyser, so the next step will be to construct the first working prototype of the emo.analyser. After user evaluation, the functionalities of the emo.analyser will be integrated into a choice-based user interface.

To conclude, this study makes two main contributions: (1) gaining more insights in: a) what coaching strategies do emotional eaters prefer; b) what triggers emotional eaters to (over)eat and at what times of the day the craving usually occur; c) emotional states at the moments of experiencing food cravings and giving in to binges, and (2) we aim at developing a general personalised eCoaching framework, enabling the optimal translation of successful behaviour change techniques, such as dialectical strategies, into persuasive coaching strategies implementable for broader target groups.

REFERENCES

[1] A. Dol, C. Bode, H. Velthuijsen, J.E.W.C. van Gemert-Pijnen, and T. van Strien, "The use of dialectical dialogues in a personalized virtual coach for obese emotional eaters: A research protocol for a field study among target group," The Ninth International Conference on eHealth, Telemedicine, and Social Medicine (eTELEMED 2017), ISSN: 2308-4359, ISBN: 978-1-61208-540-1, pp.101-104.

[2] World Health Organisation, Obesity and Overweight. October 2017. Available at: http://www.who.int/mediacentre/factsheets/fs311/en/ [retrieved: November 29, 2017]

[3] The GBD Overweight Collaborators, Health Effects of Overweight and Obesity in 195 Countries over 25 Years, 2015, The New England Journal of Medicine.

[4] T. Van Strien, C.P. Herman, and M.W. Verheijden, "Eating style, overeating and weight gain. A prospective 2-year follow-up study in a representative Dutch sample," Appetite, 12;59(3), pp. 782-789, 2012.

[5] J. Harvey-Berino et al. "Internet delivered behavioral obesity treatment," Prev Med, 08;51(2), pp. 123-128, 2010.

[6] M.J. Hutchesson et al. "eHealth interventions for the prevention and treatment of overweight and obesity in adults: a systematic review with meta-analysis," Obesity Reviews, 16, 5, pp. 376-392, 2015.

[7] G.M. Manzoni, F. Pagnini, S. Corti, E. Molinari, and G. Castelnuovo, "Internet-based behavioral interventions for obesity: An updated systematic review," Clinical Practice and Epidemiology in Mental Health, 7, pp. 19-28, 2011.

[8] M.M. Linehan, "Skills training manual for treating borderline personality disorder," New York, NY US: Guilford Press, 1993.

[9] A. Dol, O.A. Kulyk, H. Velthuijsen, J.E.W.C. van Gemert-Pijnen, and T. van Strien, "Developing a Personalised Virtual Coach 'Denk je zèlf!' for Emotional Eaters through the Design of Emotion-enriched Personas," International Journal On Advances in Life Sciences, vol 8, 3&4, 2016.

[10] H. Geissler, M. Hasenbein, S. Kanatouri, and R. Wegener, "E-Coaching: Conceptual and Empirical Findings of a Virtual Coaching Programme International," Journal of Evidence Based Coaching and Mentoring. Vol 12 (2), pp. 165 – 187, 2014.

[11] S. Kelders, O.A. Kulyk, L. van Gemert-Pijnen, and H. Oinas-Kukkonen, "Selecting effective persuasive strategies in behavior change support systems," In: Third International Workshop on Behavior Change Support Systems (BCSS 2015), In conjunction with PERSUASIVE 2015, June 3-5, Chicago, USA, 1 – 6, 2015.

[12] N. Beerlage-de Jong, A. Eikelenboom-Boskamp, A. Voss, R. Sanderman, and J. van Gemert-Pijnen, "Combining User-Centered Design with the Persuasive Systems Design Model; The Development Process of a Web-Based Registration and Monitoring System for Healthcare-Associated Infections in Nursing Homes," Int. Journal on Advances in Life Sciences, 6, pp. 262-271, 2014.

[13] O.A. Kulyk, C. den Daas, S. David, and L. van Gemert-Pijnen, "How Persuasive are Serious Games, Social Media and mHealth Technologies for Vulnerable Young Adults? Design Factors for Health Behavior and Lifestyle Change Support: Sexual Health Case," In: Proc. of the 3d Int. Workshop on Behavior Change Support Systems (BCSS 2015). In conjunction with PERSUASIVE 2015, June 3-5, Chicago, USA, 2015.

[14] J. Weizenbaum, "ELIZA--A Computer Program For the Study of Natural Language Communication Between Man and Machine," Communications of the ACM. Volume 9, Number 1, pp. 36-35, 1966.

[15] T.W. Bickmore, D. Schulman, and C. Sidner, "Automated interventions for multiple health behaviors using conversational agents," Patient education and counseling 92(2), pp. 142-148, 2013.

[16] O.A. Kulyk, H.J.A. op den Akker, R. Klaassen, and J.E.W.C. van Gemert-Pijnen, "Personalized Virtual Coaching for Lifestyle Support: Principles for Design and Evaluation," International Journal on Advances in Life Sciences, 6;3&4, pp. 300-309, 2014.

[17] A. Van Wissen, C. Vinkers, and A. van Halteren, "Developing a Virtual Coach for Chronic Patients: A User Study on the Impact of Similarity, Familiarity and Realism," Proceedings 11th International Conference, PERSUASIVE 2016 Salzburg, Austria, April 5–7, 2016.

[18] E. Hudlicka, "Virtual training and coaching of health behavior: Example from mindfulness meditation training," Patient Education and Counseling. 92, pp. 160–166, 2013.

[19] R.J. Beun, et al., "Improving Adherence in Automated e-Coaching. A case from Insomnia Therapy," Proceedings 11th International Conference, PERSUASIVE 2016 Salzburg, Austria, April 5–7, 2016.

[20] A. Lentferink et al., "Self-tracking and Persuasive eCoaching in Healthy Lifestyle Interventions: Work-in-progress Scoping Review of Key Components," Proceedings of the Fourth Int. Workshop on Behavior Change Support Systems (BCSS16), co-located with the 11th Int. Conference PERSUASIVE 2016, Salzburg, Austria, April 5, pp. 15-35, 2016.

[21] T.R. Lynch, A.L. Chapman, M.Z. Rosenthal, J.R. Kuo, and M.M. Linehan, "Mechanisms of change in dialectical behavior therapy: Theoretical and empirical observations," J Clin Psychol. 04;62(4), pp. 459-480, 2006.

[22] D.L. Safer, C.F. Telch, and E.Y. Chen, "Dialectical Behaviour Therapy for Binge Eating and Bulimia," New York: The Guilford Press, 2009.

[23] Beun, R.J., Brinkman, W.P., Fitrianie, S., Griffioen-Both, F., Horsch, C., Lancee, J., Spruit, A.G.L. Improving Adherence in Automated e-Coaching. A case from Insomnia Therapy. Proceedings 11th International Conference, PERSUASIVE 2016 Salzburg, Austria, April 5–7 (2016)

[24] H. Oinas-Kukkonen, and M. Harjumaa, "Persuasive Systems Design: Key Issues, Process Model, and System Features. Communications of the Association for Information Systems," 24(1), 28, pp. 485-500, 2009.

[25] Oinas-Kukkonen, H., A foundation for the study of behavior change support systems. Pers Ubiquit Comput. 17, 1223-1235 (2013)

[26] T. Van Strien, J.E.R. Frijters, G.P.A. Bergers, and P.B. Defares, "The Dutch Eating Behavior Questionnaire (DEBQ) for Assessment of Restrained, Emotional, and External Eating Behavior," International Journal of Eating Disorders, 5(2), pp. 295-315, 1986.

[27] The Quick Big Five Persoonlijkheidstest. Handleiding. A.A. Vermulst & J.E.M. Gerris. Leeuwarden CDC.

[28] M. Dainton, M., Zelley, E.D.: Chapter 5: Explaining theories of persuasion. Applying communication theory for professional life: A practical introduction (1st ed.) (2004) Retrieved from http://ezproxy2.utwente.nl:2180/upm-data/4985_Dainton_Chapter_5.pdf

[29] https://www.chatbots.org/chatbot/ [retrieved: November 29, 2017]

[30] "PIONIER Project," University of Groningen, The Netherlands. Available at: http://www.let.rug.nl/vannoord/alp/Alpino/ [retrieved: November 29, 2017].

[31] S. Shiffman, A.A., Stone, and M.R. Hufford, "Ecological momentary assessment," Annu Rev Clin Psychol, 4:1–3.6, 2008.

[32] I. Myin-Germeys et al., "Experience sampling research in psychopathology: opening the black box of daily life," Psychological Medicine, 39, pp. 1533-1547, 2009.

[33] Experimental Vignette Studies in Survey Research, Christiane Atzmüller and Peter M. Steiner, Methodology 2010; Vol. 6(3):128–138. DOI: 10.1027/1614-2241/a000014, 2010 Hogrefe Publishing

[34] The Use of Vignettes in Qualitative Research, Christine Barter and Emma Renold, Issue 25, 1999, Social Research Update is published quarterly by the Department of Sociology, University of Surrey, Guildford GU2 7XH, England.

[35] Finch, J. (1987) The Vignette Technique in Survey Research, Sociology, 21 (1) pp.105-114

[36] Qualtrics Available at: https://www.qualtrics.com/ [retrieved: November 29, 2017]

[37] D. Watson, L.A. Clark, and A. Tellegen, "Development and validation of brief measures of positive and negative affect: The PANAS scales," Journal of Personality and Social Psychology, Vol 54(6), pp. 1063-1070, 1988.

[38] E.J. Van Gemert-Pijnen et al. "A holistic framework to improve the uptake and impact of ehealth technologies," J Med Internet Res, vol.13, no. 4, 2011, e111.

Quantifying the Benefits of Digital Health Investments in Canada

Calculating the cumulative quality, access and productivity benefits on a national scale

Short Paper

Bobby Gheorghiu, Simon Hagens
Evaluation Services
Canada Health Infoway/Inforoute Santé du Canada
Toronto, Canada
Email: bgheorghiu@infoway-inforoute.ca
shagens@infoway-inforoute.ca

Abstract - **Capturing the benefits generated from investments in digital health is key towards demonstrating accountability to funders as well as to encouraging widespread adoption by clinicians and other health care professionals. The cumulative benefits calculation, developed by Canada Health Infoway, is a macro-level indicator trended since 2007. It represents estimated benefits accruing to various health care system stakeholders, as driven by component technologies and their associated adoption across the country. In-depth studies, validated by external experts in relevant fields, have been completed for diagnostic imaging systems, primary and ambulatory care electronic medical records, drug information systems and telehealth. The financially quantifiable aspects of each study are aggregated, trended over time, and indexed to inflation. From 2007 to 2016 benefits accrued to the Canadian health care system exceeded $19B. These benefits were driven by improvements such as clinician and clinical practice productivity; avoided health system utilization due to improved patient safety; reduced patient time and expense and fewer duplicate tests. Cumulative benefits represent the aggregated value accrued to various health system stakeholders that has been realized or could be harvested through re-organized business processes.**

Keywords-electronic health record (EHR); adoption; digital health; health system benefits; productivity; efficiency; telehealth; electronic medical record (EMR).

I. INTRODUCTION

This paper is an extended version of work published in [1]. We extend our previous work by updating data for an additional year (2016), elaborating on the studies supporting the findings, and further detailing the methods, inputs and limitations of the study. Federal and provincial/territorial governments across Canada have been investing in the creation of interoperable electronic health records (iEHRs) and other means for connecting health care providers with patient information for more than a decade [2]. While deployment progress and rates of usage vary across provinces and territories, the initiative has reached mainstream adoption by clinicians and other health care professionals. An iEHR is a secure, integrated view of a person's medical records from all systems in the network; it provides a comprehensive view of a patient's medical history. It is designed to facilitate the sharing of data across the continuum of care, health care delivery organizations, and geographical areas [3]. The potential benefits of the iEHR are substantial – improved quality of care, health system and provider efficiencies, improved access to care and use of health data to better manage the health system and facilitate research. Typically, the iEHR integrates diagnostic imaging, laboratory, and medication data, along with clinical notes, to provide a longitudinal view of a patient's clinical history. As such, it is a similar concept to that of a Health Information Exchange (HIE) [4]. Across Canada, iEHRs are at various stages of implementation and maturity and have evolved according to provincial/territorial strategies and priorities. As of 2017, each province and territory in Canada has an iEHR in place [5].

The iEHR acts as a complement to point of service applications such as EMRs in physician offices or hospital information systems. It is accessed through integration with clinical systems or through standalone, web-enabled viewers. Figure 1 illustrates the relationship between point of service applications and the iEHR infostructure (clinical data repositories in blue boxes) [6].

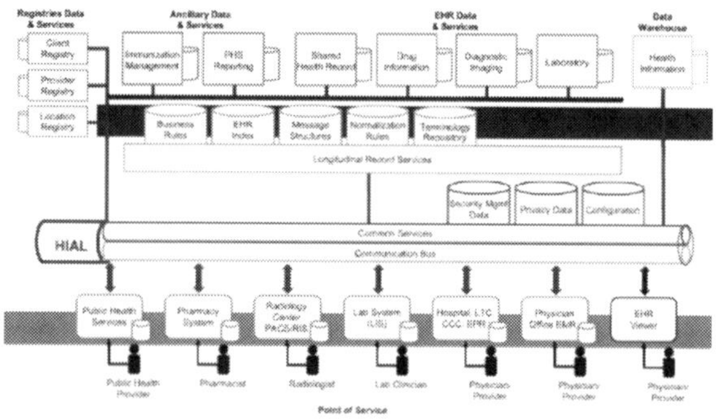

Figure 1. iEHR jurisdictional infostructure based on the Electronic Health Record Solution Blueprint.

Regular measurement of adoption and maturity for these technologies has made progress easy to follow and manage. For example, in the 2015 Commonwealth Fund Survey, 73% of all family physicians reported they do use electronic records to enter and retrieve clinical notes [7]. They form part of the 301,000 health care professionals (more than half of estimated potential users) who were accessing one or more of the following sources for patient information needed to provide care (lab, DI, drug) in 2017. Of these, more than 162,000 were actively accessing two or more of these clinical information sources [8][9]. As the trend in Figure 2 illustrates, the number of active users has accelerated greatly in the past three years as adoption of connected health information reached a critical mass.

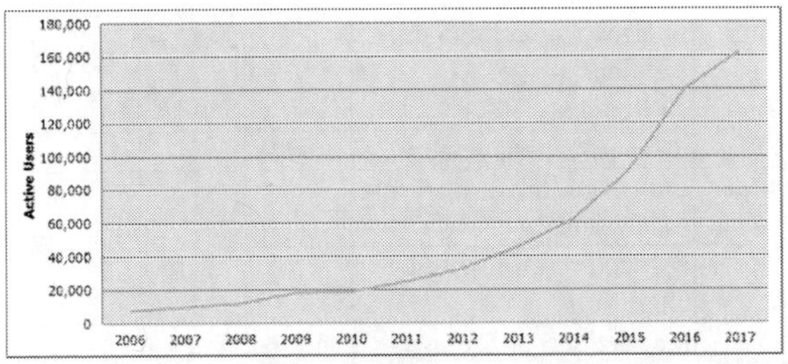

Figure 2. Trending iEHR systems use (2+ clinical domains). Active users have accessed the system a minimum of one time per month.

The rapid increase in adoption is partially attributed to the integration of iEHR data into point-of-care systems in hospitals and clinics over the last two years. Healthcare practitioners and leaders report that this integration is allowing providers to incorporate iEHR data into clinical workflows [8].

In the interest of accountability for the public funds under its management, and optimizing the value accruing from investment of those funds, Canada Health Infoway (Infoway) has developed approaches to evaluate and systematically model the estimated value of outcomes related to select digital health solutions nationally [10]. The cumulative benefits model contains estimates for benefits generated through the use of diagnostic imaging systems, drug information systems (DIS), ambulatory and primary care electronic medical records (EMR), and telehealth.

The rest of this paper is organized as follows: Section II describes the methods used for calculating cumulative benefits. Section III describes the most recent results based on data up to the end of 2016. Section IV addresses the assumptions and limitations of the model. The conclusion, Section V closes the article.

II. METHODS

The cumulative benefits calculation is driven by individual quantitative estimates obtained from pan-Canadian studies commissioned by Infoway for each of the clinical domains mentioned above. The concept of pan-Canadian studies to estimate national value was developed in 2006 to summarize results across diverse data domains, settings, evaluation methods, and time periods. The studies aim to generate estimates, which are as comprehensive as possible, peer reviewed by expert panels, and reflecting best available evidence. The cumulative benefits studied cover a subset of digital health solutions, and as such represent a portion of the value from digital health at large. The studies explored domains where federal funds directed through Infoway have been invested. A fulsome methodology, published in 2015, provides an assessment of gaps and recommendations for increasing and optimizing the use and spread of technologies in order to increase value over time [10]. Estimates are calculated in Canadian dollars realized on an annual basis and base assumptions to current contexts are applied and documented. Not all outcomes represent direct financial savings, but where possible, a value is expressed financially to allow comparison of magnitudes. Where the literature does not provide sufficient evidence to quantify the current dollar value of a specific outcome, the value is omitted from quantitative modeling. In instances where a range of estimated benefits is provided, the mid-point value was used as a base estimate. The mid-point estimate for each domain-specific benefit is highlighted in Table I and corresponds to the year in which the study was carried out. Simplicity is a core principle of the specification of the quantitative benefits model, with most discrete value estimates derived by multiplying the magnitude of outcome observed per unit x value of outcome x extent of adoption across Canada. Adoption maturity variation (e.g., functionalities used, frequency of use, etc.) is an important driver of value in digital health deployment, so the extent of the adoption used in the model must be matched to the maturity required to achieve the magnitude outcome applied.

TABLE I. YEARLY BENEFITS BY DOMAIN

	2007	2008	2009	2010	2011	2012	2013	2014	2015	2016
DI	$511	$646	$715	$724	$841	$850	$878	$888	$908	$968
DIS	$233	$369	$436	$441	$462	$487	$551	$579	$593	$865
TH	$52	$70	$95	$128	$158	$187	$266	$340	$407	$580
EMR	$131	$162	$193	$227	$270	$302	$347	$414	$419	$424
AMB-EMR	$45	$53	$59	$71	$82	$106	$141	$167	$196	$199
TOT.	$973	$1,300	$1,498	$1,590	$1,812	$1,932	$2,183	$2,387	$2,523	$3,036

Adoption is determined by examining data from surveys of clinicians and patients, usage data from digital health solutions, and operational data sets collected by Infoway's partners. Specific definitions of adoption are designed to suit distinct kinds of solutions, and trended over time [10]. Adoption is measured differently for each domain or technology. This is both due to the practicality and feasibility of collecting the data and the way in which benefits were initially modelled. Adoption metrics are applied in the model as drivers of the benefit magnitude. For example, telehealth benefits are driven by the number of clinical sessions—defined as consultations involving a clinician and a patient [11] whereas EMR benefits are driven by the number of Canadian physicians who reported using an EMR to document patient information as shown in Figure 3. EMR benefits largely accrue to physicians and their practices, and as a result, it is fitting for physician adoption to be a driving factor.

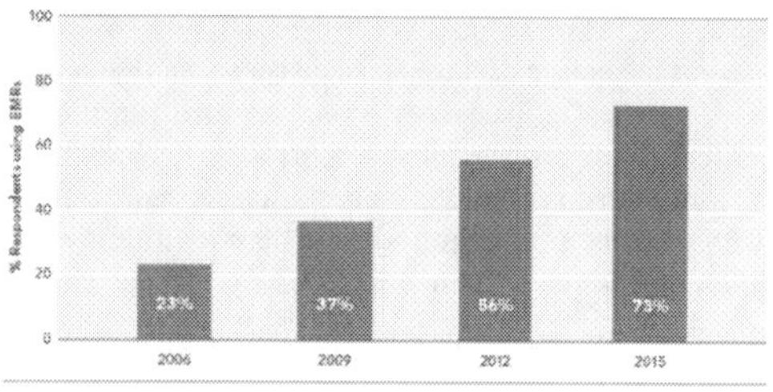

Figure 3. Family physicians in Canada reporting EMR use: Commonwealth Fund Surveys.

Conversely, telehealth benefits largely accrue to patients and the health care systems that fund patient expenses, such as the Northern Travel Grant in Ontario, which subsidizes patient and caregiver travel costs [12]. Diagnostic imaging benefits are driven by the adoption of picture archiving and communication systems (PACS) by radiologists, physician specialists, and emergency department nurses since they are the professions who would be the most likely to benefit from the workflow improvements facilitated by these systems. While other clinicians and health professionals are likely to benefit, it was preferable to be conservative in the adoption estimate, in line with Infoway's overall reporting principles [10]. Drug information system benefits are driven by a combination of the percentage of pharmacies connected to a DIS across a given province and the percentage of connected physician EMRs. This reflects the distributed nature of DIS benefits where significant portions accrue to pharmacies, physician practices, and health system funders among others.

Table II summarizes the various types of benefits measured for each respective study:

TABLE II. BREAKDOWN OF BENEFITS (DUPLICATE BENEFITS REMOVED)

		Health system use	Clinician/staff productivity	Patients
Tele-health	personal travel			$325
	subsidized travel	$158		
	ED avoidance	$3		
	hospital avoidance	$65		
DI	patient transfers	$15		
	radiologist productivity		$261	
	technologist productivity		$135	
	referring physician productivity		$175	
	duplicate exams	$-		
	film costs		$518	
DIS	community adverse drug events (ADE)	$79		
	admission ADEs	$42		
	printed/typed Script ADEs	$48		
	medication abuse	$139		
	medication compliance	$182		
	pharmacist/technician efficiency		$157	
	call-backs		$117	
	drug cost management	$88		
EMR	improved lab and DI test management		$127	
	reduced or eliminated chart pulls		$115	
	reduced lab or DI test duplicates	$136		
	reduced ADEs due to prescription legibility	$-		
AMB-EMR	avoided delays for patients			$4
	reduced or eliminated chart pulls		$95	
	reduced lab or DI test duplicates	$83		
	reduced ADEs	$-		
TOTAL ($M)	$3,067	$1,038	$1,700	$329
		34%	55%	11%

III. RESULTS

Between 2007 and 2016, over $19 billion in quantifiable benefits have accrued to various parts of the Canadian health care system as shown in Figure 4.

The diagnostic imaging study completed in 2008 documented benefits totaling an estimated $600M as a result of improved productivity for doctors and technologists (by more than 25%); improved remote reporting capabilities; improved access to care; and quicker turnaround time (by 30-40%), meaning patients get diagnosed twice as fast and are able to start treatment sooner [13].

The Generation 2 Drug Information Systems Pan-Canadian Study, completed in 2010, demonstrated benefits totaling $436M through improved safety and quality of care, fewer adverse drug events, reduced prescription abuse,

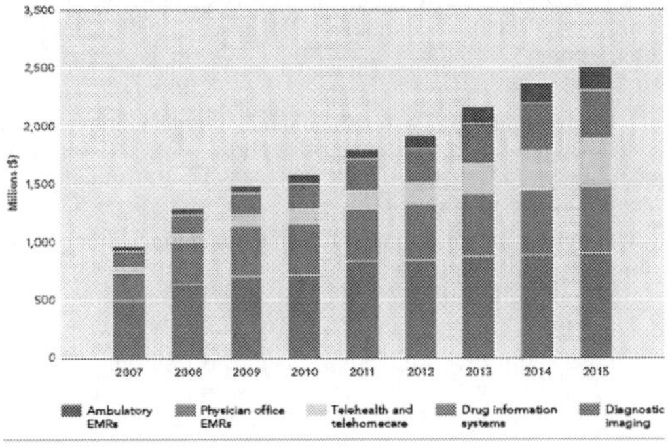

Figure 4. Cumulative benefits of investments in digital health across Canada.

increased medication compliance) along with increased provider productivity (greater pharmacist efficiency and fewer call-backs) and improved drug cost management [14].

From 2006 to 2012, the growth of EMRs in Canada resulted in cumulative efficiency and patient care benefits valued at more than $1.3 billion--$800 million in administrative efficiencies and $584 million in health system level benefits, such as reduced duplicate tests and adverse drug events [15].

Lastly, use of EMRs in ambulatory care clinics across the country has resulted in benefits estimated at $196 million in 2015 due to clinic efficiencies (improved chart management, reduced duplicate tests and reduced transcription costs); increased clinic capacity; fewer emergency department visits and hospitalizations as a result of adverse drug events; and avoided costs for patients due to avoidable delays in care [16].

Coloured cells in Table I represent base calculations in the years when each respective study was completed. Benefit estimates for previous and subsequent years are dependent on the changes in adoption of each technology by health care professionals and/or practices in each year as compared to the base years. Each year's estimate is adjusted for inflation according to the Statistics Canada Consumer Price Index for Health Care [17]. In 2016, the estimated value to the overall healthcare system in aggregate exceeded $3B. These benefits accrue to various stakeholders within the system, as shown in Figure 5. The largest piece (54%) accrues to health delivery organizations (HDO) such as hospitals, clinics, pharmacies along with their respective clinicians and other staff. The next largest piece (36%) accrues to health system funders such as ministries of health and other government organizations as a result of reductions in reimbursable health system utilization and subsidies. Lastly, 10% of quantified benefits accrue to patients and their families largely through time saved in accessing care and avoided travel.

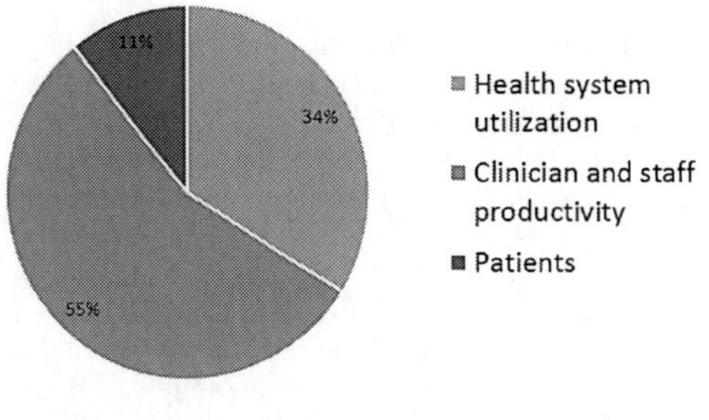

Figure 5. Distribution of benefits across the Canadian health care system

It is expected that future pan-Canadian studies, examining incremental benefits related to other components of the iEHR such as patient access to health information and/or electronic prescribing, will expand the model according to the same overarching methods described above.

IV. ASSUMPTIONS AND LIMITATIONS

The pan-Canadian studies completed to date, which drive the cumulative benefits calculation, were completed independently, over a number of years (2008 to 2015). As such, benefits calculated as part of one study, may also be reflected in another. For example, the Generation 2 Drug Information Systems study tracked patient safety benefits related to printing out prescriptions from physician EMRs—this benefit was also included in the EMR pan-Canadian study. These benefits were included only once when aggregating totals in order to eliminate double counting. However, due to differing methodologies in each study, double counting cannot be eliminated completely.

Significant limitations were noted in the availability of data upon which to model and quantify certain kinds of benefits. For example, anecdotal evidence exists to support the idea that productivity and quality improvements in healthcare delivery would result in benefits for patients in terms of time saved, avoided losses in productivity, and improved convenience among others; however, insufficient data was available upon which to make evidence-based estimates. As a consequence some types of benefits are understated in the cumulative analysis, such as value to patients, which is estimated at only 10% of the total. Another important example relates to improved compliance with quality of care guidelines. Physicians with EMRs are more proactive in caring for their patients, but the long-term financial implications of these improvements could not be modelled or included with quantitative estimates. As such, these benefits are largely addressed only qualitatively in the pan-Canadian studies.

While it would be desirable to calculate return on investment (ROI), benefits and costs accrue to multiple stakeholders, making these calculations complicated, and best assessed on a stakeholder by stakeholder basis. Furthermore, while some stakeholders may see quantifiable benefits from a digital health investment, others may have to bear additional costs. Such an example can be found with PACS where productivity gains for physicians may increase access for patients, but result in additional costs to the health system since more clinical procedures and/or patient visits can be completed with the same inputs. Nonetheless, this is a useful analysis at a macroeconomic level since it can demonstrate the positive effects of investing in digital health. An important limitation of this analysis and digital health progress as a whole is the gap between value created and hard savings. This limitation was also identified in analysis conducted by the RAND Corporation. In 2005, researchers estimated that rapid adoption of health information technology could save the United States more than $81 billion annually [18]. In 2012 RAND revisited the topic and found mixed results and growing health expenditures. The factors that limited achievement of value were "sluggish adoption of health IT systems, coupled with the choice of systems that are neither interoperable nor easy to use; and the failure of health care providers and institutions to reengineer care processes to reap the full benefits of health IT" [19]. In the Canadian context, similar issues apply, but steps are being taken to address these. Adoption incentives, focus on interoperability, and attention to change management have been important for driving mature adoption of solutions. The pan-Canadian studies apply metrics of adoption and maturity as drivers to model benefits, so estimated value reflects actual changes in clinical practice. In contrast, however, there are aspects of reengineering care processes and more broadly, health system organization and financing that are not addressed. For example, time savings that accrue to providers who operate on a fee for service basis are gains to the specific providers, rather than the health system. Reductions in ER and hospital utilization can add capacity, but for hard-savings, Canadian health systems must look at how different sectors of the system are financed. As digital health becomes the norm, health systems leaders are increasingly aware of the need to be proactive in the harvesting of benefits.

V. CONCLUSION

The cumulative benefits calculation has been and continues to be a useful tool for demonstrating the benefits of digital health investments for the purpose of accountability to funders and taxpayers. In addition, it is a useful tool to persuade clinicians and other health care professional to adopt new technologies, and to encourage partners, such as jurisdictions and health care provider organizations to continue to invest in digital health. Its main limitation is that while the total figure represents aggregate value to various stakeholders, it does not represent actual dollar savings, since other workflow, policy, and personnel changes may need to be implemented in order to fully harvest the benefits.

As connected health information continues to become more accessible across a number of clinical domains, it is contributing to the continued improvement of quality, access and productivity for patients, health care professionals, and the health system overall.

REFERENCES

[1] B. Gheorghiu and S. Hagens, "Cumulative benefits of digital health investments in Canada," eTELEMED 2017, The Ninth International Conference on eHealth, Telemedicine, and Social Medicine. Mar. 2017; ISBN: 978-1-61208-540-1.

[2] S. Tharmalingam, S. Hagens, and J. Zelmer. "The value of connected health information: perceptions of electronic health record users in Canada." 2016. BMC medical informatics and decision making, 16(1), 93.

[3] Canada Health Infoway. (2016). *Interoperable EHRs*. [Online]. Available from: https://www.infoway-inforoute.ca/index.php/programs-services/investment-programs/interoperable-ehr. 2017.02.16.

[4] Healthcare Information and Management Systems Society, HIMSS dictionary of healthcare information technology terms, acronyms and organizations, Third Edition, 2013.

[5] Canada Health Infoway. (2017). *Year in review, 2016-2017*. [Online]. Available from: https://www.infoway-inforoute.ca/en/component/edocman/3350-annual-report-2016-2017/view-document. 2017.08.05.

[6] Canada Health Infoway. (2006). *EHRS blueprint: An interoperable EHR framework*. [Online]. Available from: https://www.infoway-inforoute.ca/en/component/edocman/391-ehrs-blueprint-v2-full/. 2017.08.11.

[7] R. Osborn et al., "Primary care physicians in ten countries report challenges caring for patients with complex health needs." Health Affairs 34.12 (2015): 2104-2112.

[8] B. Gheorghiu and S. Hagens, "Measuring interoperable EHR adoption and maturity: a Canadian example," BMC Med Inform Decis Mak. Jan. 2016; doi:10.1186/s12911-016-0247-x.

[9] Canada Health Infoway. (2016). *A conversation about digital health, Annual Report 2015-2016*. [Online]. Available from: https://www.infoway-inforoute.ca/en/component/edocman/3098-annualreport-2015-2016/view-document. 2017.02.01.

[10] S. Hagens, J. Zelmer, C. Frazer, B. Gheorghiu, and C. Leaver, "Valuing National Effects of Digital Health Investments: An Applied Method," Information Technology and Communications in Health (ITCH 2015), Feb. 2015, pp.165-169

[11] Canada Health Infoway. (2011). *Telehealth Benefits and Adoption: Connecting People and Providers*. [Online]. Available from: https://www.infoway-inforoute.ca/en/component/edocman/333-telehealth-benefits-and-adoption-connecting-people-and-providers-full/view-document?Itemid=101. 2017.02.11.

[12] Ontario Ministry of Health and Long-Term Care. (2017). *Northern Health Travel Grants*. [Online]. Available from: http://www.health.gov.on.ca/en/public/publications/ohip/northern.aspx. 2017.07.08.

[13] Canada Health Infoway. (2008). *Diagnostic Imaging Benefits Evaluation Final Report*. [Online]. Available from: https://www.infoway-inforoute.ca/en/component/edocman/329-diagnostic-imaging-benefits-evaluation-report-full. 2017.07.08.

[14] Deloitte (2010). *National Impacts of Generation 2 Drug Information Systems Technical Report*. [Online]. Available from: https://www.infoway-inforoute.ca/en/component/edocman/331-national-impact-of-generation-2-drug-information-systems-technical-report-full/. 2017.07.09.

[15] PricewaterhouseCoopers (PwC). (2013). *The emerging benefits of electronic medical record use in community-based care*. [Online]. Available from: https://www.infoway-inforoute.ca/en/component/edocman/1224-the-emerging-benefits-of-electronic-medical-record-use-in-community-based-care-full-report/. 2017.07.15.

[16] PricewaterhouseCoopers (PwC). (2016). *The emerging benefits of EMR use in ambulatory care in Canada: Benefits Evaluation Study*. [Online]. Available from: https://www.infoway-inforoute.ca/en/component/edocman/3025-the-emerging-benefits-of-emr-use-in-ambulatory-care-in-canada-full-report/. 2017.07.15.

[17] Statistics Canada. *Consumer price index, health and personal care, by province*. [Online]. Available from: http://www.statcan.gc.ca/tables-tableaux/sum-som/l01/cst01/econ161a-eng.htm. 2017.02.18.

[18] R. Hillestad et al., (2005). Can electronic medical record systems transform health care? Potential health benefits, savings, and costs. Health affairs, 24(5), 1103-1117.

[19] A. Kellermann and S. Jones., "What it will take to achieve the as-yet-unfulfilled promises of health information technology." Health affairs 32.1 (2013): 63-68.

A Method for Architectural Inclusive Design: the Case of Users Experiencing Down Syndrome

Clémentine Schelings, Catherine Elsen
LUCID-ULg
University of Liège
Liège, Belgium
e-mail: {clementine.schelings; catherine.elsen}@ulg.ac.be

Abstract—**This paper develops an in-situ methodology to help architects insure better inclusion of people with Down syndrome all along preliminary phases of the architectural design process, and eventually to the designed space. This methodology first offers architects some design keys in regard of how people with Down syndrome interact with two types of spaces: their personal dwellings and some completely unknown spaces. The methodology then unfolds towards more pro-active inclusion of the participants thanks to playful expression of their feelings and perceptions. This paper discusses how this methodology relates to inclusive and universal principles, useful to design smart environments, be they ICT-enabled or not. This paper closes on prevalent models of disability in architecture and how they articulate with the model of "architectural handicap".**

Keywords-disability; Down syndrome; inclusive design; universal design; methodological framework.

I. INTRODUCTION

This paper, an extended version of a previous, shorter publication presented at the conference Smart Accessibility 2017 [1], tackles the challenge of disability inclusion to architecture, disability considered here as a temporary or permanent condition likely to show up at any time of everyone's life. Statistically speaking, disability concerns 15% of the European population, i.e., more than 80 millions individuals [2]. Among them, only 20% are disabled from birth, while 80% will experience impairment later in life, as a result of an accident, an illness, ageing or a more temporary condition such as pregnancy [3]. We are therefore all concerned with disability, whatever our current situation.

Designers are yet struggling with the inclusion of disabled people, given the challenge of end-users' inclusion into the design process in general, the variety of disabilities and the variety of adaptations those disabilities require on both spatial and functional levels. In architectural design, and in Belgium more specifically, norms about persons with reduced mobility (PRM) constitute one of the few frameworks available to help designers integrate the needs of people who use a wheelchair or the needs of blind people. This regulation, yet, does not take into account cognitive impairments (nor hearing loss) that are thus generally neglected during the architectural design process. Likewise, in ICT related fields, cognitive impairments seem to be less often considered than visual or hearing loss impairments [4].

Consequently, this paper aims at offering some concrete design methodology to architects confronted to the needs of people with cognitive disabilities, and more specifically people with Down syndrome. The paper will first aim at studying the impact of architecture on the spatial perception of people with cognitive disabilities. In-situ observations of participants evolving through various spaces will provide some useful design keys in that regard. The methodology will then be expanded in order to include those users into a more active encounter with architecture, providing architects with fruitful information about how people with Down syndrome experience space on a more multi-sensory level.

This paper is structured in five more sections. In Sections II and III, literature review and the resulting research questions are presented. Section IV details the methodology developed in order to conduct the observations. Section V describes the obtained results, presented in four subsections: space perception of participants with Down syndrome (Subsection A), representativeness of those results (Subsection B), handicapping situations (Subsection C) and methodological recommendations for inclusion of Down syndrome in architectural design (Subsection D). Section VI closes on a theoretical discussion considering prevalent models of inclusion and disability in architecture and how these models should be revised in order to consider people with Down syndrome's sensitiveness as opportunity, rather than threat, to the architectural design process. Some insights built in this paper might be relevant for universal/inclusive design in ICT related fields.

II. STATE OF THE ART

Since the modernist era, a distance has been created between the architect and the end-user, the latter being considered as some abstract, hypothetical subject required to adjust him-/herself to the designed building [5]. Phenomenologists such as Juhani Pallasmaa argue that this "modernist reductionism", i.e., this way of conceiving some architectural artifact that "standard" users will encounter mainly in a visual way, has globally impoverished both the professional praxis and the architectural experience [6][7]. In order to re-instate a more inclusive, multi-sensorial approach of architecture (both in its design and experience), some researchers and practitioners have developed since the 70's some hands-on, participative methodologies. One has to observe that these methodologies, although being well-meaning initiatives, often remain too distant from architectural practice, or too superficial in their implementation, and therefore still offer considerable room for improvement [8]. Indeed, participation methodologies have generally been implemented either as a way to collect end-users' post-building assessment [9], or as a way to secure users' acceptance at the very late design phases [10], rather than a tool helpful to take into account users' needs and perceptions as soon as the early phases of the design process [11]. This lack of effective upstream frameworks makes it very difficult for architects to integrate laypeople into the design process, and makes it even harder when considering a wider range of end-users, including people with disabilities.

This literature review section is structured into four subsections, starting with some observations about the current state of architectural practice, followed with the challenge of including disabled users into the design process, continuing with the specificities of people with Down syndrome and ending with an overview of how disability could be considered as an expertise, instead as an obstacle to design and creativity.

A. About architecture, end-users and their well-being

We highlight here two main observations from architectural state of the art: the uni-sensoriality and the environmental impact of current architectural design.

1) Uni-sensoriality

As observed by several phenomenologists, most of current architectural design projects are almost exclusively visually designed [7]: architects strive to impress with attractive and pleasing graphical designs, and often invest less efforts into the integration of users' various ways of experiencing space [12][13]. Architecture, authors argue, consequently suffers some kind of uni-sensoriality hegemony [6]. Yet architecture intrinsically is a multi-sensorial experience, since "*qualities of space, matter and scale are measured equally by the eye, ear, nose, skin, skeleton and muscle* [6 (p.41)]." In fact, all our senses interact together and complement each other in order to shape our understanding of the world, and particularly the spatial environment. For instance, the human mind unconsciously associates visual shapes with odors and emotions because "*the most persistent memory of any space is often its smell* [6 (p.54)]." Some authors associate this uni-sensoriality hegemony with an impoverishment of architectural experience and praxis, and retrace its roots to the modernist era where architecture, according to their critical historical analysis, has been reduced to the sole consequence of visual expression and experience, therefore neglecting the other perceptual senses and consequently deviating from the users' multi-sensorial realities [6][14][15].

This focus on sight is even sometimes considered as an "handicap" for the architects themselves [12], as such uni-sensorial approach does not only reduce the human capabilities but also the design opportunities and qualities. The whole body should therefore be considered as a complex "thinking organ" [16] that combines multiple intelligences such as visual, haptic and kinesthetic ones. Taking into account the complete scope of the human mind and body, phenomenologists argue, aims at avoiding superficial and dehumanized architectural design [6] and at proposing more sensitive and creative projects.

2) Environmental impact

Theories of environmental psychology and healing environments suggest that the architectural environment influences the wellbeing, considering architecture either as a factor having a positive (curative architecture) or a negative (disabling architecture) impact on the emotional and physical experience [17][18].

According to those theories, architects themselves sometimes accidentally generate discomfort and dissatisfaction feelings for the end-users of their buildings, simply by being unaware of the consequences their "disabling design" might have on people [18]. Most of the time those "architectural disablements" are completely independent of any medical disability end-users might themselves experience, and are rather more fundamentally related to building impediments such as, most commonly, ill-designed stairs or uncomfortable, confined spaces [13]. This is particularly true for a parent, for instance, who might experience difficulties when pushing a stroller through some heavy doors or narrow hallways [13]. As a result, the henceforth "architecturally disabled people" either manage to overcome the obstacles but only with effort and frustration, or simply are prevented to use the building.

B. About the integration of disabled users

Most prevalent approaches characterize disability as a constraint for both designers and users. Considering disability, and the norms associated to it, as obstacles to their creativity [19], most architects rarely expand their effort of integration beyond the simplest form of a user, such as the "*average, six-foot-tall, 20-years-old male, with perfect vision and a good grip* [20 (p. 60.7)]." As a consequence, users experiencing disability (either as a permanent or temporary condition) rarely see their specific needs and perceptions taken into account. This denial of diversity finds its roots as much as the uni-sensoriality hegemony in the modernist quest for economical and aesthetic design, considering only invariant body proportions [21] and predictable users' needs [22].

When relating to existing norms and regulations regarding disabled people, architects are moreover only informed about a limited variety of disabilities, not even considering variations within the same disability. Reference documents and standard procedures, for length and clarity, indeed generally tackle a limited range of bodily (in)capabilities. As a result, over-simplification of their interpretation causes shortcuts from "people with disabilities" to "people using a wheelchair" [21]. One has to observe that the main studied disabilities are actually motor impairments and blindness, while cognitive impairments are more rarely addressed, except for autism that has been explored [23][24]. The resulting recommendations and designs are thus never perfectly adapted to the users with cognitive disability, whose needs and necessities are misunderstood and excluded [23].

C. *About people with Down syndrome*

Among the variety of cognitive disabilities one could design for, this paper focuses on Down syndrome, a genetic anomaly linked to the presence of a third chromosome on the 21st pair. This is not an illness per se, but rather a physical and cognitive state of the person expressed by several characteristics like physical appearance, medical fragility, cognitive disability, peaceable temperament and short life expectancy [25][26][27][28].

Among all those specificities shared by people with Down syndrome, some have a direct impact on the way they experiment architectural environments and spaces. First, their small size influence their space perception as their field-of-view is lower than average [26]. Second, because of their premature ageing [27], people with Down syndrome often develop early motor impairments and try to limit their moves, this behavior being yet sometimes counteracted by their curious and playful nature. Third, due to their lack of concentration [28], the presence of a disruptive element can generate trouble and monopolize so much of their attention that they completely ignore other environmental or spatial factors. They are also known to give much importance to rituals and habits in their daily lives, in order to limit the occurrence of such disruptive events. Yet, people with Down syndrome can show high adaptive capacity: after a while, they tend to accept all kinds of situations, even uncomfortable ones. In fact, their affability [28] disposes them to go beyond the drawbacks of one situation and to eventually get used to it.

Eventually, the cognitive disability brings two specific features regarding spatial apprehension. People with Down syndrome are first of all particularly receptive to their multi-sensory experience of space [28]. Similarly to people with autism spectrum disorders [29], people with Down syndrome moreover present a remarkable hypersensitivity and a particular spatial perception that induces spontaneous emotions and instinctive reactions towards some space. This spontaneity and peculiar way to experience space, revealing some of its specificities (qualities and/or defects), could potentially enrich assessment of any building for instance. By doing so, "*turning disability experience into expertise in assessing building accessibility* [30 (p. 144)]" or in designing multi-sensorial spaces [7], a concept that has essentially focused on motor and visual impairments until now, could open towards other types of disability such as Down syndrome.

D. *About considering disability as an expertise*

Encouraging architects to question and reinstate users' multi-sensoriality and sensitivity into their work, a few researchers propose to interact with disabled people and to integrate their perceptions as soon as early stages of the design process [12][31]. The disability is then considered as an opportunity, both for architects who develop new ideas and for disabled people who take part into a process from which they are usually excluded. In this case, disabled people are considered as experts and become a real source of creativity for designers [31]. The literature documents two ways to integrate the disabled people's expertise in the design process.

1) *Disability as specific expertise for disabled users*

In this case, disabled users are experts of their own ways of experiencing space and the generated results directly benefit people with similar disabilities. For instance, Tufvesson's and Tufvesson's study compares pupils with autism, hyperactivity and Down syndrome and provides some design strategies and environmental factors with positive (e.g., individual desks) or negative (e.g., low acoustical insulation) effects on the work atmosphere [29]. Studying children with learning disabilities thus primarily provides solutions for disabled children.

2) *Disability as general expertise for all users*

Another approach takes advantage of the expertise of disabled users either to design innovative projects addressing the needs of a larger audience, or to tackle additional challenges of the architectural field in general. A good example of that approach is Penezic and Rogina design of the "Glass House 2001 for a blind man" [32]. Given that transparency and reflection are both unperceivable properties for people with visual impairment, these architects developed technologies so that any user would experiment the building in an equivalent way. As a consequence, the walls are constituted of two layers of glass between which a fluid circulates at different temperatures and speeds, this way mobilizing two alternative perceptual senses, the touch and the hearing. Besides being particularly innovative and inclusive, this ingenious system solves two recurrent problems of glass buildings, i.e., acoustics and thermal insulation.

III. RESEARCH QUESTIONS

Considering people with Down syndrome peculiar multi-sensoriality and hypersensitivity, we suggest that their spatial experience should be valued as specific expertise and considered as a way to reinstate multi-sensorial qualities into architectural design, and to limit "handicapping situations" for end-users. To this end, a methodological framework has to be developed in response to their specific needs and necessities.

This methodological framework will be nurtured by the answers to the following three research questions:

- How do people with Down syndrome perceive space at a multi-sensory level?
- Which "handicapping situations" do people with Down syndrome face?
- How to leverage Down syndrome's specificities and expertise for architectural design?

IV. Methodology

To answer those research questions, we build an original methodology of in-situ observation and in-situ interaction with disabled participants, inspired by theories of inclusive design and Nijs' and Heylighen's own research methodology [30].

A. *Inclusive design methodologies*

Inclusive design theory relies on two main principles, i.e., (i) considering the users' and designers' complementarity given their respective specific knowledge and expertise [33], and (ii) re-integrating the users' experiences, emotions and reactions in order to design sensitive architecture ensuring their wellbeing [16].

Several methodologies share the idea that every user has a "*potential equal contribution to the design outcomes*" [34 (p. 524)]. For instance, methodologies of co-design and participative design endeavor to overcome the historical denial of end-users by giving them a real voice in the decision-making process.

Nonetheless, this objective is still considered ambitious because it threatens two widespread conceptions: on the one hand, the professional architect seen as the sole legitimate author of architecture, and, on the other hand, the user seen as an obedient spectator, contemplating the building as a piece of art [22]. As a matter of fact, most architects reject unpredictable and creative usages as well as any behavior that deviates from the original design of the building, starting from the premise that "*art is the product of individual creativity*" [22 (p. 22)].

When such methodologies are nevertheless implemented, a gap might subsist between architect's and end-users' habituses, i.e., between specialized knowledge of architectural practice and unfamiliarity with the architectural design process, that might generate misunderstandings or even conflicts [35]. End-users, although given a voice, might indeed feel disempowered when confronted to the unknown field of architecture, especially when they are asked to participate without being provided with any kind of support or formation [36].

The methodology developed here, inspired by the inclusive design theory and principles, therefore grant special care to the architect/investigator posture and to the support provided to the participants all along its in-situ implementation.

B. *Nijs' and Heylighen's research mehodology*

Given the reluctance towards users' inclusion in general and the limits of current frameworks, Nijs and Heylighen have developed a specific methodology that considers disabled people as experts of their own peculiar way of experiencing spatiality and architecture [30]. Through several cases studies, these researchers invited disabled people to visit a building and to discuss their own experience verbally, thanks to different keywords suggested by the researchers.

The novelty of this approach, compared to classic accessibility assessments, is the creation of four groups, one for each type of disability (motor, visual, auditory and cognitive impairments). The methodology is the same for each group: participants were asked to follow a predefined route across the building while one of them, designated "research assistant", had to fill out a specific evaluation sheet for the group. The collected data are mainly the formalized transcript of participants' oral comments when moving freely in the rooms, and the expression of their feelings about spatial quality and accessibility. Supplemented by some pictures, the resulting evaluation report attests to the group's specific point of view and is then collectively debated with the other groups in order to write the final common report. Eventually, the participants' experience, identified problems and proposed solutions could be communicated to building practitioners and designers.

We believe this methodology works perfectly for three groups out of four, but is only half adapted to people with cognitive impairments, and more specifically to persons with Down syndrome. Indeed, those users undoubtedly show potential expertise in visiting different architectural spaces and in assessing their multi-sensorial qualities, but capturing that information by written and oral comments and debates seems unsuitable. While Subsection IV.C will develop how we implemented our specific methodology, Section V will thus come back on how and why our methodology had to be adapted in regard of Nijs' and Heylighen's one, given the communication difficulties of people with Down syndrome.

C. *Implemented methodology*

The methodology has been developed iteratively, on basis of two campaigns of observation. This section will first present the test observation and then the main observation, both corresponding to different observation techniques and different roles of the observer.

1) Test observation

A good observation is characterized by a good preparation and requires a well-trained researcher [37]. A first step of our methodology, particularly suggested to architects who would be willing to implement it, is to conduct some test observations in order to practice and to highlight the major elements requiring particular attention, i.e., the main themes of a future observation grid [38].

Before starting our main observations, we thus chose to test and fine-tune a first version of our observation criteria. This first testing phase was conducted in the context of a guided tour organized by a museum for a group of people with cognitive disabilities, including people with Down syndrome. During the visit, the researcher took the role of a complete observer, meaning that the participants had no clue they were observed, and used the "fly on the wall" observation technique [39]. As the name suggests, the observer remains discreet and collects information without interacting with the observed people. The main advantages are the participants' spontaneity and the possibility to quickly gather data from several persons at the same time.

This first observation expanded our observation grid with some attention points such as emotions, disruptive elements and events as well as particular attention to personal vocabulary. Moreover, this visit of the museum initiated the idea of a playful interaction with the participants, which would go beyond the "fly on the wall" discretion of the researcher but would also offer richer insights in terms of participants' perceptions and emotions. The test observation eventually made us particularly aware of the impact an unknown place, such as a museum, might have on the participants.

2) Main observation

Initiating the main observation, we firstly proceeded to the selection of the participants affected by Down syndrome. Six participants were eventually chosen among the residents of a Belgian non-profit association welcoming adults with cognitive disabilities, and specifically intended to develop residents' artistic skills. Those participants were chosen on the basis of several criteria such as the sex (to ensure gender parity), the housing type (in order to compare the participants' experience in terms of living with family or living permanently in the residence) or the severity of their disability and the impact it could have on their capability to express their experiences and feelings (Tab. I).

Secondly, we conducted two phases of in-situ observations: first the visit of the residents' own dwellings and later the discovery of a public building, a local town hall unknown by the participants. Those two observation sequences were video-recorded to ease post comparative analysis. The goal here was to compare the spatial perceptions of people with Down syndrome when confronted to familiar vs. unknown spaces. This choice, inspired by the test observation, was additionally confirmed by two different aspects of the state of the art. On the one hand, the review of Down syndrome's characteristics had informed us about the importance of daily routines and spatial memory [28], and we assumed we would more easily observe the impact of such factors inside some usual environment, i.e., inside the participants' private dwelling. On the other hand, research about disabling architecture shows that handicapping situations generally occur in public buildings [13], which confirmed our choice of unusual environment.

At the beginning of the visit of each dwelling, we set up a discussion table in order to collect some basic information such as, for instance, the resident's age or favorite room(s). This stage also helped us create a climate of confidence with the participant and his or her referee (family member or close relative), invited to join the whole observation in order to ease communication and interaction. We then organized a playful activity, operationalized with the help of two psychologists, which consisted in visiting the resident's three preferred rooms and interviewing him or her about his or her felt experience thanks to illustrated cards.

This combination of observation and interview methods, close to the "shadowing" technique, enables the researcher to follow a person in his or her daily activities while asking him or her some questions to complete the observed information [40]. Within this framework, the researcher takes over the role of observer-as-participant, i.e., he or she spends more time observing than participating, while always clarifying the scientific objectives. This role has several benefits: it is especially adapted for short interviews, it enables real-time filling of observation grids and it ensures transparency of the research goals towards the observed subjects [41]. However, given the brevity of each session (40 minutes in average), a mutual misunderstanding can occur between the observer and the observed person. Hence the need to quickly build confidence [41], which could mainly be achieved thanks to the presence of the participant's relatives.

TABLE I. DEMOGRAPHIC PROFILE OF THE PARTICIPANTS HIGHLIGHTING SOME ADDITIONAL SPECIFICITIES

#	Selection criteria				
	Gender	***Age***	***Housing type***	***Cognitive specificity***	***Mobility specificity***
1	female	25	family house	/	artificial hip
2	female	48	residence	/	slower motion
3	male	27	family house	verbalizes through onomatopoeias	/
4	male	36	family house	/	/
5	male	27	residence	/	/
6	male	49	residence	/	/

The methodology implemented during the visit of the town hall was rather similar: a few days later, we invited the same six participants to visit three rooms of the town hall, this time chosen by the researcher in order to compare each participant's reactions. The visit of those three selected rooms was made individually. In the meantime, the five other participants were guided by a social worker for a recreational photo activity in order to capture their experience when they were visiting the town hall on their own, in the absence of the observer.

The last step of our methodology was an artistic activity later organized in the drafting room of the day center. Participants were asked to draw or paint the buildings we visited the previous days. They could choose to sketch their private dwelling and/or the town hall from the inside and/or the outside, delivering their personal interpretation of those spaces. The drawings produced, as well as the pictures taken by the residents are an additional means of expression completing or confirming the information collected during the individual visits.

V. RESULTS

The four next sections will present the results of the in-situ observations, starting with factors impacting space perception of participants with Down syndrome and representativeness of those results, following with observations of "handicapping situations" and ending with some methodological recommendations.

A. Space Perception

During the two observation phases, five main phenomena have been observed.

Firstly, the people with Down syndrome who took part to this study all experienced some difficulties in identifying the limits between spaces that were not clearly delineated by a physical boundary. In the town hall for instance, the reception and entrance halls were separated by a simple inner bay frame (Fig. 1), but the participants designated those two spaces as one single room. When asked to walk around the reception hall, they indeed systematically travelled both halls, obviously confused by the proximity of two sub-spaces whose functions and boundaries were insufficiently distinct. Similarly in the case of private dwelling, one participant walked around the living room when asked to delineate the kitchen.

Secondly, and in contrast with the previous point, people with Down syndrome who took part to this study paid particular attention to the privacy of a space and how this sense of privacy could delineate one space from another. During the visits of their dwellings, the participants have always chosen their own bedroom as their favorite room, which underlines their need to have a personal, private space available. This characteristic could also be observed while experiencing the public building, especially when some residents felt the need to be alone and left in search of some smaller, more comfortable and/or less traveled space to retreat to for some time. In the case of their private spaces (their rooms), privacy did, in spite of its intangible nature, build some boundary between two subspaces. This phenomenon was specifically observed in a bedroom shared by two residents who never crossed the invisible line dividing the room into two individual and appropriated zones.

Thirdly, the participants demonstrated a particular attraction for light, bay windows, illuminated objects and surfaces. This characteristic was observed repeatedly, particularly when participants were asked to point to their favorite object within a room. One of them, for instance, showed us his stereo, occupying a special spot on the windowsill of his bedroom, which was particularly well lit. This importance of natural light is moreover clearly illustrated by the participants through their drawings of the visited buildings. Most of the time, their representations were rather simple but they always involved drawings of the windows, including specific details such as frames, shadows and/or glazing (Fig. 2).

Fourthly, our observations revealed the great importance of material landmarks in the everyday-life of the participants, especially in regard of their day-to-day rituals and habits. Those well-known elements, which could be objects, pieces of furniture or even a specific building material (e.g., local brown stone), were reassuring to them especially because they reminded them of aspects of their daily life and environments. In one of the residences, we visited a living room that had just been rearranged and refurnished. Inside this living room, social workers had left a small wooden table (Fig. 3) greatly appreciated by the participants because it had been crafted by one of the residents. This small table, placed there as a landmark of the previous space configuration, greatly facilitated the occupants' appropriation of this new way of organizing the room. The presence of this recognizable piece of furniture helped the acceptance of a new situation otherwise potentially disturbing.

Fifthly, spatial perception of the participants was strongly impacted by their personal preferences and areas of interest. During the visits, the participants spontaneously went towards objects or pieces of furniture referring to one of their passions or to an episode of their personal history. For instance, one of the participants' interest for photography influenced his moves in the town hall, his path being essentially oriented towards specific photo frames.

Besides those five keys of space perception, we have observed two additional mechanisms engaged in different settings: first the visuo-spatial memory participants developed in regard of everyday spaces, and second the multi-sensoriality participants deployed especially in unknown spaces. Those two additional phenomena confirm our initial hypothesis stating that comparing familiar and unfamiliar places would reveal different behaviors regarding space perception.

Figure 1. Reception and entrance halls separated by an inner bay frame.

Figure 2. One participant's drawing of the town hall and its bay windows.

Figure 3. Wooden table in the living room of one residence: the reassuring landmark easing the space re-organisation and appropriation.

When interviewed inside their dwellings, the residents generally looked beyond the current situation and appealed to their memory to describe the space as they generally experience it, rather than describing it in regard of its specificities at the time of observation. For instance, one participant stated that the living room was a place where "*it was dark*" while it was a bright middle of the afternoon at the time. The participant described the room as he usually perceives it in situation of most frequent use, i.e., when he watches TV in the evening, this way appealing to his visuo-spatial memory instead of his instant capacities of observation. Another resident displayed the same memory capacity when he showed us the living room, which was being renovated at the time of the visit. Although all the furniture was stacked in the corner of the room, the participant told us which armchair was his favorite and, even though the room was very quiet, he explained us it was an animated place where he usually spent time with his friend. Compared with other rooms visited with this participant, we observed that he took more time to answer our questions, probably because he was looking for responses in his memories and could/would not rely on his feelings on the fly.

In the town hall, moreover, participants largely mobilized their five senses to experience space. For example, they relied on their hearing to determine the level of activity of the rooms: one participant said that the entrance hall was "*here, quiet, everything is quiet*" because we were alone in the room, while another one later found the space "*animated*" because several employees were present at the time. Some participants also appealed to kinesthesia and explored almost every inch of the room in order to appreciate the spatial quality of the room before answering the questions of the observer. We observed that multi-sensoriality was generally only engaged during the discovery phases of a new space or a potentially disturbing environment.

B. Representativeness

As mentioned earlier, the results were gathered from a rather limited sample of only six people with Down Syndrome. However, such a qualitative study provides representative results on the condition that the saturation criterion is respected.

The saturation criterion, originating from the social sciences, is "*the point at which no new information or themes are observed in the data* [42 (p.59)]," which means that the size of the sample may be limited to 'n' people if the person 'n+1' reveals no new essential data compared to the person 'n'. This theory thus argues that the size of the sample, i.e., the number of observed people, should not be defined by a theoretical number fixed in advance, but should emerge from the research field [43]. The goal here is not to reach any statistical representativeness, but rather to "*reflect at best the possible variety of the testimonies* [44 (p. 58)]" through results built with observed data and confirmed by the forthcoming ones. In this prospect, social scientists use comparative analysis and follow an iterative process, constantly adapting their theory until recurrences reveal some clues of saturation [45].

Practically speaking, we made sure that the saturation point was reached by building a table structuring comparative analysis (Table II), in which the six space perception phenomena and the data provided by each of the six participants (both for their dwelling [D] and for the town hall [T]) were crossed. The idea was to check whether the results were observed in any situation, i.e., whether the saturation point was potentially reached, or if some exceptions remained.

The Table II, presenting a simplified version of this comparative analysis table, uses five symbols:

- "V" represents a phenomenon effectively observed in the dwelling of the resident;
- "O" shows a phenomenon effectively observed in the town hall;
- "X" corresponds to the absence of use of visuo-spatial memory in unknown places;
- "/" refers to the absence of multi-sensorial experience in familiar places;
- "!" indicates an exception, i.e., an unobserved phenomenon where we would have logically expected it to occur.

The theory of saturation is not necessarily opposed to the existence of exceptions and "*the singular fact, the negative cases, the misunderstood or enigmatic phenomena, although they are essential, do not call into question this saturation* [44 (p. 305)]." In our case, all exceptions find a rather simple explanation and some of them even share the same one. In Table II, we thus numbered from 1 to 5 each type of exceptions, for which we will provide below specific explanation.

TABLE II. COMPARATIVE ANALYSIS AND REPRESENTATIVENESS OF THE RESULTS REGARDING THE PARTICIPANTS' SPACE PERCEPTION

Space perception	Participants											
	1		*2*		*3*		*4*		*5*		*6*	
	D	***T***	***D***	***T***	***D***	***T***	***D***	***T***	***D***	***T***	***D***	***T***
Limits	V	!1	V	O	V	!1	!2	!1	V	O	!3	O
Privacy	V	O	V	O	V	!1	V	!1	V	O	V	O
Light	V	O	V	O	V	O	!4	O	V	O	!5	!5
Material landmark	V	O	V	O	V	O	V	O	V	O	V	O
Individual preferences	V	O	V	O	V	O	V	O	V	O	V	O
Visuo-spatial memory	V	X	V	X	V	X	V	X	V	X	V	X
Multisensoriality	/	O	/	O	/	O	/	O	/	O	/	O

- D: in the dwelling
- T: in the town hall
- V & O: observed phenomena
- X & /: logically not observed phenomena
- !1, !2, !3, !4 & !5: exceptions

In the first case (exceptions !1), two phenomena regarding space perception were not observed in the town hall: troubles with perceiving limits and search for an intimate space or a hiding place, that occurred only for half of the participants. For the other half such phenomenon didn't occur, either because some of them refused to participate when asked to walk around the room or because they were so tired that they remained seated in each room we visited. Under these circumstances, we cannot conclude whether they had difficulties with limits or not, as well as whether they felt the need for privacy or not.

In the second case (exception !2), the third resident did not manifest any problem with the identification of limits in his dwelling. In fact, all the spaces of his private house were clearly delimited by walls and doors, leaving no place for spatial confusion.

In the third case (exception !3), the sixth participant didn't expressed problems with limits in the residence where he stayed. All spaces were not necessarily physically delimited, but the levels of privacy generated some invisible but perceptible boundaries between them. Therefore, the resident clearly differentiated one functional area (e.g., TV lounge) from another (e.g., game corner) even if they were in the same large room.

In the fourth case (exception !4), the visit of the fourth participant's dwelling took place at the end of the day, when it was already dark outside. Obviously, we did not observe anything regarding natural light, but that might have been the case at a different time of the day.

In the fifth case (exceptions !5), the sixth participant did not show any interest for light and even preferred dark spaces, which can seem really surprising. Nevertheless, his appeal for darkness is linked to his character and personal history (see Section VI).

Finally, each phenomenon has been observed at least once for each participant, except for the two dark grey boxes of Table II. Consequently, identical phenomena were observed for each participant, without any new trend when increasing the sample in size. The results therefore remain representative and respect the saturation criterion, allowing the involvement of small samples of participants in such methodology. However, the compared results also highlight that there is a variety of realities even inside the same disability. Building on this observation, our study suggests reconsidering the prevalent models of disability (see Section VI).

C. *Handicapping situations*

During the visits, we observed the two most common handicapping situations as defined by Goldsmith, i.e., confined spaces and stairs [13], but also a third one, which is the height of the furniture. Every time, the disabling situation is caused by an incompatibility between the spatial configuration and the users' characteristics.

In the residences, participants shared their rooms with other people and their personal space is reduced to a small area. As a result, nearly all the rooms become common places and there is no specific area really respectful of privacy, which is yet very important for the participants as our observations revealed (see Section V.A). As a social worker told us, when one resident is arguing with others, he or she cannot take refuge in a quiet place to calm down. In order to limit these incidents and to respect the participants' need for privacy, individual bedrooms are recommended as well as several little shelters in different rooms of the residence.

During the visits, we also observed that some participants with minor mobility impairments (e.g., a slight limp) hardly came up and down the stairs. In fact, besides the effort it takes them to climb the stairs, the residents were afraid of loosing their balance and falling down. Steps without risers are even scarier and, in her house, one resident had no choice but to slide down the spiral stairs on her back. In our view, the absence of risers is mainly justified by a search for aesthetics but it generates collateral damages like discomfort and fear, which could be easily limited with riders or even avoided with single stories buildings.

As previously mentioned, participants with Down syndrome are small-sized and, thus, the usual height of furniture is generally not suitable for them. For instance, one participant had difficulty in sitting on the kitchen barstools and lost his balance because his feet did not touch the ground (nor the footrest). Similarly, social workers made some adjustments when providing lower shelves to store the objects that the residents use on a daily basis.

In those three handicapping situations, design choices can have architecture-independent consequences such as discomfort, promiscuity or risk of falling because they are not adapted to the residents' needs. In all cases, simple solutions can be found as soon as architects anticipate the users' specificities, an anticipation that requires a specific methodology.

D. Methodological Recommendations

Compared to traditional approaches, inclusive design requires a meeting with the end-users as soon as preliminary design phases. Critics may argue that considering end-users from the beginning is time-consuming, but on the contrary we believe it saves time in the long term and avoids calling into question projects that are about to be built. Moreover, our observations demonstrate that there are possibilities to get to know the users within a short time (about one hour), which should be considered as an investment rather than a waste of time. However, in order to make the most of the meetings with the users, the designer should be open-minded towards their requests and suggestions, keeping in mind that end-users are experts of their own needs and ways of living. Yet, one has to observe that it is not natural for end-users to take part in the design process by documenting their daily behaviors and expressing their feelings about space. Therefore, we suggest developing renewed forms of interaction (other than simple conversations), such as for instance narrative inquiries, scenario-based approaches, or, as for participants with Down syndrome, playful activities.

The following paragraphs summarize adaptations made to Nijs and Heylighen's methodology [30] in order to make it more suitable to the specificities of people with cognitive disability (for which oral expression and debate, for instance, can be difficult).

To begin with, the importance of the referee (family member, close relative or educator) was made clear during the first phases of "discussion tables" we added to the methodology: this actor, acting as mediator between the observer and the observed person, played a crucial role in decoding both stakeholders' words, intentions and behaviors and in ensuring their mutual understanding and trust. In one particular case, the presence of the participant's parents turned out to be essential to "translate" his personal vocabulary mainly composed of onomatopoeias. Furthermore, the referee generally took pleasure recounting some anecdotes, which complemented the resident's comments and behaviors and contributed to faster reach the saturation point.

Expression of feelings and perceptual spatial experiences were moreover greatly facilitated by the use of four cards illustrated with cartoony human faces, each featuring one of the most widespread human primary emotions (happiness, sadness, nervousness and fear). These cards, chosen with the help of a psychologist specialized in assisting people with Down syndrome, were voluntary simple (free of superfluous details) and limited in their number in order to help participants express their feelings as accurately, as well as simply, as possible. Participants were nevertheless free to combine several pictures to enrich their answers if necessary. Those cards, as suggested by Chase, adequately complement the content usually collected through narrative inquiry [46]. One important preliminary step, when presenting these cards for the first time, was to proceed to the emotions' recognition, i.e., to align our understanding to what the cards meant in the eyes of the participants. For instance, one resident had identified the card of the scared figure as a person "*who winced*", and this definition was therefore used for the rest of those observations. Those cards proved really useful to interact with the participants once on the field, and could efficiently replace the keywords used by Nijs and Heylighen [30] when interacting with people experiencing difficulties with verbal expression.

From an organizational perspective, we visited each room in two phases: first, we started interviewing the participant, and then we let him or her walk around the room. During the visit of one dwelling, one of the residents at first refused to sit and to answer our questions. We had to wait until he stopped moving before obtaining a single answer. Organizing the intervention in several, distinct and repeatable phases thus allowed us to progressively channel the resident's attention on our questions. We moreover observed that interviewing each participant separately proved particularly important to avoid participants influencing each other: at one point of the town hall visit, all six participants started to interact about the space and the influence of one of them was clearly at the disadvantage of self-expression.

Eventually, considering additional means of expression, such as photography or drawing for instance, proved very useful to complete some participants' comments.

VI. Discussion

Our in-situ observations contribute to an adapted methodology and to design keys useful for architects willing to include people with Down syndrome (their specific needs, their specific ways of experiencing space) into preliminary phases of their design processes. Since the results presented here are issued from six participants only, the findings should not be generalized to a larger group. As Kinnaer, Baumers and Heylighen underline in their research about autism, individual preferences play an important role for the perception and appreciation of certain spaces and should not be dismissed [24]. This has proven also true for people with Down syndrome, as one of the participants distinguished from the five others by his particular appeal for dark spaces. In this case, the participant considered his own bedroom, indeed rather dark, as his personal shelter of privacy, a space where he could freely unleash his emotions. He therefore associated dark spaces to this personal space, a protective cocoon where he could express himself untroubled. Designers willing to replicate the suggested adapted methodology should therefore apply the saturation criterion [42] as a way to capture both specific and shared spatial perceptions.

Down syndrome, as any other cognitive disabilities, consequently ought to be considered as a complex condition, characterized by a variety of realities unfortunately confined to a rather limited global medical model [23]. Yet, current theoretical and practical disability frameworks hardly take into account this variability. In general, end-users' integration is hardly achieved, and reveals even more problematic for architects considering regulations about people with disability as an obstacle to their creativity rather than as a support to their design. Moreover, those regulations have the tendency to reduce the user to a single, « representative » profile: even the architectural norms applied to the inclusion of PRM tend to dismiss personal specificities one wheelchair user can develop in regard of another. Theories such as Universal design, on the other hand, intend to transform architecture into some universal product including the diversity of needs of all potential users [47]. To put it another way, universal architecture goes beyond "design for special needs" and proposes a built environment usable by every user without exception and under all circumstances [18]. Such Universal architecture, by doing so, might even reduce the model of the user and his/her uses, as each Universal user potentially accumulates the incapacities of a larger diversity of users, the designed object being consequently reduced to its lowest common possible use [48].

This research is therefore rather in favor of the inclusive model, taking into account the specificities of users and considering them, as much as possible, as creative input. Even thought providing a real voice to end-users, inclusive design still goes with some limitations, such as becoming a danger for disabled users' autonomy. In some cases, the search for "absolute well-being" leads to the design of spaces perfectly adapted to the users' needs, at the risk of creating overprotected and aseptic places in contradiction with users' personal development [23]. In the case of people with cognitive disabilities, there is a need to find a delicate balance between users' autonomy and prosthetic architecture, i.e., between stimulating users' empowerment [23] and providing adjustments that ease daily life and counteract impairments [49]. Regarding the height of furniture for instance, most used objects by participants with Down syndrome could be stored on the lowest shelf boards accessible without any help, while a stepladder could be used to reach the highest objects when necessary. In this way, a safe and comfortable environment would be provided on a day-to-day basis and the residents would be encouraged to make punctual efforts and to look for alternative solutions when required.

We argue that the methodology developed in this paper, favoring playfulness rather than simple consultation of the end-users, might potentially help architects conducting in-situ research and gaining knowledge about how specific groups of people with Down syndrome interact with architecture. In our case, a playful approach was favored following the advice of a child psychologist and given the pretty childlike nature of people with Down syndrome, but the methodology should be adapted according to the users' profile and the type of disability concerned. Therefore, this methodology is characterized by its flexibility and proposes a non-rigid reflexive framework for designers. Furthermore, participants, considered as experts of their own disability and their own specific ways of experiencing space, might in this way contribute to architectural projects more prone to benefit the greatest number of users. As much as hypersensitivity [29], people with Down syndrome's specific ways to apprehend an architectural space, for instance through higher multi-sensoriality, could equip designers in their perception of end-users' needs. Whereas universal design aims at the lowest common denominator, inclusive design, we argue, provides more diversified avenues for design exploration.

Including participants with Down syndrome as soon as preliminary phases of the architectural design process, and specifically empowering them with a certain expertise, moreover suggests a possible evolution of current models of handicap in architecture. Disability has originally been considered the result of a medical condition, therefore building the "medical model" of disability in architecture. This model, focusing exclusively on disability as an illness together with its symptoms, nurtured a hygienist design of specialized institutions. At that time, architects had no responsibility at all regarding exclusion of people with physical and/or mental impairments, whose specificities were identified as the cause of handicap creation [8]. Furthermore, disabled users were supposed to adapt themselves to the built environment whatever their individual characteristics [18]. Later, a social model of disability in architecture rather focused on the human being rather than on the mere "patient", and integrated notions such as "*origin, milieu, education, profession, economical position and social status* [50 (p. 11)], quoted by [51 (p. 19)]" to the design of adapted spaces. This social model, as a consequence, informed the design of healing environments outside the institutionalized boundaries of the hospitals and proposed living environments "*accommodating people with a social framework and, thus, supporting residents in developing their identity* [51 (p. 24)]." Following this model, some architects became highly conscious about environmental impact on our perceptions; and even managed to use this architectural externality as a design key. For instance, the Maggie's Centers are well known for their spatial capacity of becoming psychological and moral support for cancer-sick patients, which is enhanced by a welcoming and convivial architecture [52].

Following our observations, we would advocate a third model of disability, i.e., architecture considered as a potentially disabling factor. This model, as an extension of the social model, would "*focus on individuality, difference (instead of commonality), experience and giving voice to people* [51 (p. 25)]," while redefining the role of both architecture and the architects.

This concept, introduced by Goldsmith in the context of a research focusing on motor and visual impairments [13], states that architecture can constitute a proper physical barrier as much for disabled users than for people with temporary limited mobility (injured or pregnant person for instance). This "architectural handicap" therefore translates into an uncomfortable and constraining situation for the user, caused by the lack of consideration or anticipation from the designer that would not, or could not take into account the specificities of a larger group of potential users [18].

We argue this notion of architectural handicap extends to any type of disability, including cognitive ones, as well as any type of design field, including ICT-related ones. In the case of people with Down syndrome, our results suggest that architecture sometimes not only constitutes some physical barrier to one's mobility, but also a psychological barrier (e.g., loss of autonomy, fear of falling or lack of privacy). Unclearly delineated spaces, for instance, can generate loss of reference points, misunderstanding of sub-functions and consequently loss of autonomy and social exclusion.

Architecture and architects therefore have a crucial role to play in terms of avoiding such handicapping situations: the design keys and methodology proposed in this paper offer support to architects who wish to deal with this new responsibility.

CONCLUSION AND FUTURE WORK

This paper develops a methodology to approach Down syndrome in architectural design, in line with inclusive design theories. The originality of this methodology lies in its early integration of participants and its playfulness, enabling to go beyond simple consultation with end-users and to value the disability experience as an expertise.

The methodology and design keys suggested in this paper may be suitable to other user profiles, such as people bearers of another cognitive impairment, seniors or children who share some characteristics with people with Down syndrome. However, this methodology should not be applied as it stands for all user profiles, but should remain flexible and adaptable according to their specificities.

Our research also highlights the limits of the current normative frameworks. Nonetheless, the actual lack of consideration for people with cognitive impairment compared with other disabilities, like motor impairment, demonstrates the benefits of such a norm. Since a strict regulatory framework would not be an adequate solution, this paper rather paves the way for a toolbox for designers, encouraging them to take into account people with cognitive disability and suggesting them some interaction techniques to reach this goal.

No longer considering disability as a threat or obstacle for architectural design, this work rather suggests that people with Down syndrome experience space with some specific sensitiveness. This sensitiveness could be leveraged as a source of creativity for the designer ("disability as opportunity"), while architecture could be considered as a potentially handicapping factor for the user ("architectural handicap").

ACKNOWLEDGMENTS

The authors would like to thank all who contributed to this research, in particular the team and residents of "Les Hautes Ardennes" center (Vielsalm, Belgium). The authors also thank Ms. Henry, psychologist at "Cité de l'Espoir" (Andrimont, Belgium), for her advices in carefully choosing the content of the illustrated cards, and Ms. Stiévenart, child psychologist and researcher (University of Liège, Belgium), for her suggestions for the operationalization of the playful activity.

REFERENCES

[1] C. Schelings and C. Elsen, "Inclusion of Down syndrome in architectural design: towards a methodology," in Smart Accessibility 2017: The Second International Conference on Universal Accessibility in the Internet of Things and Smart Environments, L. Smirek, D. Marghitu and K. S. Fuglerud, Eds. Nice : IARIA, pp. 20-25, 2017.

[2] AViQ Agence pour une Vie de Qualité. *How many people are disabled in Belgium, in Wallonia?* [Online]. Available from: https://www.aviq.be/handicap/questions/infos_conseils/statistiques.html 2017.11.13

[3] AViQ Agence pour une Vie de Qualité. *The disabled workers? Workers!* [Online]. Available from: https://www.aviq.be/handicap/pdf/documentation/publications/emploi/Brochure-Travailleurs-handicapes.pdf 2017.11.13

[4] A. Bai, H. C. Mork, T. Halbach, K. S. Fuglerud, W. Leister, and T. Schulz, "A review of universal design in ambient intelligence environments," in Smart Accessibility 2016: The First International Conference on Universal Accessibility in the Internet of Things and Smart Environments, L. Smirek and K. S. Fuglerud, Eds. Nice: IARIA, pp. 6-11, 2016.

[5] F. Choay, Urban planning, utopias and realities An anthology. Paris: Editions du Seuil, 1965.

[6] J. Pallasmaa, The eyes of the skin. Chichester: John Wiley, 2005.

[7] J. Herssens and A. Heylighen, "Haptics and vision in architecture: designing for more senses," Conference Sensory Urbanism, Flâneur Press, Jan. 2008, pp. 102-112, ISBN: 978-0-95599-060-1.

[8] Y. Lee, "Design participation tactics: the challenges and new roles for designers in the co-design process," CoDesign, vol. 4 (1), pp. 31-50, 2008, doi: 10.1080/15710880701875613.

[9] F. Y. Y. Ling and C. L. L. Chong, "Design-and-build contractors' service quality in public projects in Singapore," Building and Environment, vol. 40 (6), pp. 815-823, Jul. 2004, doi: 10.1016/j.buildenv.2004.07.017.

[10] R. Luck, "Learning to talk to users in participatory design situations," Design Studies, vol. 28 (3), pp. 217-242, May 2007, doi: 10.1016/j.destud.2007.02.002.

[11] M. Annemans, E. Karanastasi, and A. Heylighen, "From designing for the patient to designing for a person," in Inclusive designing Joining usability, accessibility, and inclusion, P. Langdon, J. Lazar, A. Heylighen, and H. Dong, Eds. Cambridge: Springer, pp. 189-200, 2014.

[12] A. Heylighen, P. Devlieger, and M. Strickfaden, "Design expertise as disability and vice versa," Communicating (by) Design, Chalmers University of Technology, Hogeschool voor Wetenschap & Kunst, School of Architecture Sint-Lucas, Apr. 2009, pp. 227-235, ISBN: 978-9-08132-380-2.

[13] S. Goldsmith, Designing for the disabled The new paradigm. New-York: Routledge, 2012.

[14] A. Dalhin, On architecture, aesthetic experience and the embodied mind. Seven Essays. Stockholm: Royal Institute of Technology, School of Architecture, KTH, 2002.

[15] M. Merleau-Ponty, Phenomenology of perception. London and New York: Routledge, 2002.

[16] M. Ferreira, D. Cabral de Mello, and J. Duarte, "Embodied emotions: a phenomenological approach to computation to explore empathy through architecture," Digital Physicality, Proceedings of the 30th eCAADe Conference, vol. 2, Czech Technical University in Prague, Faculty of Architecture, Sep. 2012, pp. 599-604.

[17] V. Van der Linden, M. Annemans, and A. Heylighen, "You'd want an energy from a building: user experience of healing environment in a Maggie's cancer caring centre," Proceedings of the 3rd European Conference on Design4Health, Design Society, Jul. 2015, pp. 1-9, ISBN: 978-1-84387-385-3.

[18] H. Froyen, E. Verdonck, D. De Meester, and A. Heylighen, "Documenting handicap situations and eliminations through universal design patterns," Australasian Medical Journal, vol. 1 (12), pp. 199-203, 2009, doi:10.4066/AMJ.2009.158.

[19] G. Nijs, P. Vermeersch, P. Devlieger, and A. Heylighen, "Extending the dialogue between design(ers) and disabled use(rs): from conversation to embodied skill," in DS 60:

Proceedings of Design 2010, the 11th International Design Conference, D. Marjanovic, M. Storga, N. Pavkovic, and N. Bojcetic, Eds. Dubrovnik: Design society, pp. 1817-1826, 2010.

[20] V. Fletcher, "A neighborhood fit for people," in Universal design handbook, W. Preiser and E. Ostroff, Eds. New York: McGraw-Hill, pp. 60.1–60.15, 2001.

[21] R. Imrie, "Responding to the design needs of disabled people," Journal of Urban Design, vol. 5 (2), pp. 199-219, 2000, doi: 10.1080/713683959.

[22] J. Hill, Actions of architecture Achitects and creative users. London: Routledge, 2003.

[23] K. McAllister and B. Maguire, "Design considerations for the autism spectrum disorder-friendly key stage 1 classroom," Support for Learning, vol. 27 (3), pp. 103-112, Aug. 2012, doi:10.1111/j.1467-9604.2012.01525.x.

[24] M. Kinnaer, S. Baumers, and A. Heylighen, "How do people with autism (like to) live?," in Inclusive designing Joining usability, accessibility, and inclusion, P. Langdon, J. Lazar, A. Heylighen, and H. Dong, Eds. Cambridge: Springer, pp. 175-185, 2014.

[25] RT21 Regroupement pour la trisomie 21. *Qu'est-ce que la trisomie 21?* [Online]. Available from: http://trisomie.qc.ca/apropos/quest-ce-que-la-trisomie-21/#en_detail 2017.11.13

[26] S. Thérien. *Synthesis on Down syndrome* [Online]. Available from: http://www.asdet21.org/synthese.php 2017.11.13

[27] R. Bonjean, "File: people with Down syndrome increasing age Editorial," Le journal de l'apem t-21, vol. 11, pp. 1-2, 2012.

[28] J. A. Rondal, "Neurobehavioural specificity in Down syndrome," in Congrès international neuropsychologie et comportement de la personne trisomique 21, Junglisnter: Trisomie 21 Lëtzebuerg a.s.b.l., pp. 18-25 , 2008.

[29] C. Tufvesson and J. Tufvesson, "The building process as a tool towards an all-inclusive school. A Swedish example focusing on children with defined concentration difficulties such as ADHD, autism and Down's syndrome," Journal of Housing and the Built Environment, vol. 24 (1), pp. 47-66, Apr. 2009, doi:10.1007/s10901-008-9129-6.

[30] G. Nijs and A. Heylighen, "Turning disability experience into expertise in assessing building accessibility: a contribution to articulating disability epistemology," Alter, vol. 9 (2), pp. 144-156, Apr. 2015, doi:10.1016/j.alter.2014.12.001.

[31] J. Herssens and A. Heylighen, "Haptic architecture becomes architectural hap," Ergonomics for a future, Proceedings of the 39th Annual Congres of the Nordic Ergonomic Society (NES), NES, Oct. 2007, ISBN: 978-9-19715-212-9.

[32] P&R Penezić and Rogina. *Tokyo works* [Online]. Available from: https://www.behance.net/gallery/14244251/Tokyo-Works 2017. 11.13

[33] A. Heylighen and M. Bianchin, "How does inclusive design relate to good design? Designing as a deliberative enterprise," Design Studies, vol. 34 (1), pp. 93-110, Jan. 2013, doi:10.1016/j.destud.2012.05.002.

[34] R. Luck, "Dialogue in participatory design," Design Studies, vol. 24 (6), pp. 523-535, Nov. 2003, doi: 10.1016/S0142-694X(03)00040-1.

[35] J. P. S. Siva and K. London, "Habitus shock: a model for architect-client relationships on house projects based on sociological and psychological perspectives," in CIB W096 2009: Proccedings of the CIB W096 Architectural management, C.-T. Tzeng, S. Emmitt, and M. Prins, Eds. Taiwan: National Cheng Kung University, pp. 209-220, 2017.

[36] E. Bjögvinsson, P. Ehn, and P.-A. Hillgren, "Design things and design thinking: contemporary participatory design challenges," Design Issues, vol. 28 (3), pp. 101-116, Summer 2012, doi: 10.1162/DESI_a_00165.

[37] L. Baker, "Observation: a complex research method," Library Trends, vol. 55 (1), pp. 171-189, Summer 2006, doi: 10.1353/lib.2006.0045.

[38] S. Martineau, "In-situ observation: challenges, opportunities and limits," Recherches qualitatives, special issue (2), pp. 5-17, 2005.

[39] E. Matthew-Daniel and R. Kamper. *Fly on the wall Get to know what your users want with this low-cost method for collecting valid customer data* [Online]. Available from: https://www.ibm.com/developerworks/library/us-fly/index.html 2017.11.13

[40] S. McDonald, "Studying actions in context: a qualitative shadowing method for organizational research," Qualitative Research, vol. 5 (4), pp. 455-473, Nov. 2005, doi:10.1177/1468794105056923.

[41] R. L. Gold, "Roles in sociological field observations," Social Forces, vol. 36 (3), pp. 217-223, Mar. 1958, doi:10.2307/2573808.

[42] G. Guest, A. Bunce, and L. Johnson, "How many interviews are enough ? An experiment with data saturation and variability," Field Methods, vol. 18 (1), pp. 59-82, Feb. 2006, doi: 10.1177/1525822X05279903.

[43] M. Mason, "Sample size and saturation in PhD studies using qualitative interviews," Forum Qualitative Sozialforschung / Forum: Qualitative Social Research, vol. 11 (3), art. 8, Aug. 2010, doi: 10.17169/fqs-11.3.1428.

[44] A. Cornet, The life experience of female magistrates in Frecnh-speaking Belgium (doctoral dissertation). Gilly: Université de Liège, 2011.

[45] L. Savoie-Zajc, "How to build a scientifically valid sample?" Recherches qualitatives, special issue (5), pp. 99-111, May 2006, ISSN 1715-8702.

[46] S. E. Chase, "Narrative inquiry Still a field in the making," in The sage handbook of qualitative research 4, N. K. Denzin & Y. S. Lincoln, Eds. Oaks: Sage, pp. 421-434, 2011.

[47] R. L. Mace, "Universal design in housing," Assistive Technology, vol. 10 (1), pp. 21-28, Oct. 2010, doi: 10.1080/10400435.1998.10131957.

[48] M. Winance, "Universal design and the challenge of diversity. Reflections on the principles of universal design, based on empirical research of people's mobility," Disability and Rehabilitation, vol. 36 (16), pp. 1334-1343, Jul. 2014, doi: 10.3109/09638288.2014.936564.

[49] P. Dehan, Habitat of the elderly. From adapted dwelling to Ehpad, USLD and Alzheimer units. Paris: Editions Le Moniteur, 2007.

[50] N. Mens and C. Wagenaar, The architecture of the elderly. Building for housing and care. Rotterdam: NAi Uitgevers, 2009.

[51] I. Van Steenwinckel, Offering architects insights into living with dementia Three case studies on orientation in space-time-identity (doctoral dissertation). Kampenhout: KU Leuven, 2015.

[52] C. Jenks and E. Heathcote, The architecture of hope. London: Frances Lincoln, 2010.

Energy-Aware Systems for Improving the Well-Being of Older People by Reducing Their Energy Consumption

Jon Robinson, Kevin Lee, and Raheem Yousaf

Department of Computing & Technology
Nottingham Trent University
Nottingham, UK
email:jon.robinson@ntu.ac.uk
email:kevin.lee@ntu.ac.uk
email:raheem.yousaf022012@my.ntu.ac.uk

Kofi Appiah

Department of Computing
Sheffield Hallam University
Sheffield, UK
email:k.e.appiah@shu.ac.uk

***Abstract*—Fuel poverty is becoming a problem amongst the older community in the UK. To help reduce the anxiety that fuel poverty places on older members of the community, this paper will first address why such systems are necessary before introducing a system and various interfaces for engaging and promoting better energy usage. Key areas of the proposed prototype will be discussed which focuses on a recommender and behavioural change system which enables older people to improve their energy footprint through energy-aware systems. Using systems to help reduce fuel poverty will invariably improve their general well-being. Results show how this technology can be accepted and act as an enabler in improving the overall well-being of older people as well as other system considerations. In addition, a number of subsequent phases of the project will be detailed which will discuss a longer test duration, an analysis of the data harvested and future directions.**

***Keywords*- energy-efficiency; energy-awareness; intelligent sensor systems; recommender systems; well-being; internet of things; big data.**

I. Introduction

This paper is based on [1] and provides more detail regarding the prototype system and results as first presented at the 2017 International Conference on eHealth, Telemedicine, and Social Medicine.

At present, older people within the community face a host of technological and perceptual problems, which can inhibit their interaction and well-being while using information and communications technology (ICT) solutions to improve their lives. This is especially relevant when considering Internet of Things (IoT) based systems to improve their well-being within their home environment [17], dealing with heating and reducing costs. Fuel poverty is a key societal problem that impacts on large numbers of the older community within the UK [2] as older people tend to fall into the low-income range due to low pensions and rising energy costs. The worry and anxiety concerning their limited financial resources results in 20,000 – 25,000 deaths in the UK [4]. This is due to reducing their dependency on heating their homes as they are not financially secure enough to afford or pay for their monthly household heating or cooling bills [5]. The use of technology to help older people reduce costs is constrained by their fear and distrust of technology. To enable older people to accept ICT solutions, the UK government has a number of initiatives in place to help reduce this fear (e.g., digital by default [3]), which aims to provide them with skills to use modern technology. However, figures show that high numbers of individuals are still susceptible to fuel poverty which is not only reducing their well-being but also putting their lives at risk [4].

Compounding the issue of fuel poverty, older members of the community also fear the use of technology which can impact on the successful deployment and adoption of technology to address the issue of fuel poverty. Other issues relating to the acceptance of technology and the security of personal information also factor into the distrust of ICT, which is there to help improve their well-being.

In this work a prototype system is introduced to show how unobtrusive and intelligent technology can help reduce the energy consumption of older people and how this can impact on their behaviour by adopting more energy conscious patterns through a real-time responsive recommender system. Therefore, this work will examine whether an *intelligent recommender system can reduce the energy consumption of older people by predicting and modifying their daily patterns and behaviours so that they minimise their energy needs*. The first phase of this work is to address the aforementioned requirement and expand on this by considering large scale data collection for more in-depth analysis and system considerations that need to be taken into account.

The rest of this paper is structured as follows: Section II provides an insight into existing systems; Section III discusses the design considerations that need to be taken into account when considering energy-awareness as a whole; Section IV introduces our proposed system and its key components; Section V discusses user interaction issues and factors that can deter engagement; Section VI provides initial results; Section VII provides results gathered from the second phase of system development; Section VIII discusses what will be considered in the next phases;

Section IX provides a short discussion and finally, the conclusion is presented in Section X.

II. Related Work

Reducing the energy consumption of individuals has been widely tackled by a variety of commercial and academic systems. An approach taken by some commercial systems is to view each room as a separate entity and give contextual information regarding the state of the room. Some systems require user intervention in the control of the ambient energy consumption. For example, with the *Hive* system [11], to adjust or improve the heating within a room, a separate mobile based application is required which can lead to problems when considering older people. Although additional sensors and smart plugs are available to provide the ability to turn on and off devices [22], these still need to have some form of user intervention. Voice control or a mobile phone application is required to govern their operation. However, this forces the user to actively monitor, respond and modify temperatures rather than allow the system to proactively shape the energy use of the home. IFTTT (IF This, Then That) [12] is another system which allows IoT integration within the home by providing users with a simple rule-based approach for governing the operation of smart devices. Devices are linked to a web based system which can then control the devices within the home. Systems like this would fail to address the needs and anxiety encountered when older people interact with technology [6] due to mobile technology and applications not being accessible to all older people which also increases their fear of technology. Another similar system is the Honeywell *evoHome* system [13] which provides users with contextual information about the room temperature and provides them with a digital thermostat to control ambient temperatures. The *Heat Genius* system [14] is another commercial product which uses motion control to determine whether heating is required in a particular room. This system also provides the ability to learn and adjust room temperatures based on the life-style of the users. However, these commercial systems do not address the issue of reducing energy consumption as a whole but instead focus on one specific problem i.e., heating.

Smart energy meters have been used extensively in the home and in some cases have targeted low-income and older users. In [16], a number of smart energy systems and the interactions with users are discussed. Here, users used the *Duet* and *Trio* devices in a bid to monitor their real-time energy consumption over a one year period. Smart home and assistive technology can also play an important part in monitoring and adapting the environment to best suit an older person. Kim et al. [15] discuss the *U-Health* system for monitoring and supporting the older person within their home. Wireless sensor technologies embedded within appliances in the home allow for the collection and mining of information regarding the habits of the older user, as well as providing decision-making capabilities to allow the system to adapt to the users' needs.

In [23] Granados et al. proposed an IoT architecture for healthcare applications based on an energy-efficient and resilient gateway design taking advantage of some characteristics of many home and hospital environments. Their architecture focuses on enabling wide-spread data collection from medical and structural sensors. Similarly, Rahmani et al. [24] presented a fog computing-based solution to enhance different characteristics of IoT architectures used for healthcare applications in terms of energy-efficiency, performance, reliability, and interoperability. Their fog assisted system was targeted at monitoring patients with acute illnesses.

Leandro et al. [26] proposed a smart architecture for In-Home Healthcare, based on the use of IoT for smart and individualised monitoring of elderly patients in their homes. The system has two units, a sensory unit for capturing images and a decision unit for inferring the user's health from the images.

Zhou et al. [25] presented an architecture and a set of functional modules for a smart home energy management system, with the aim of implementing an optimal demand dispatch system. The system relies on the smart grid paradigm to collect real-time electricity consumption data from in-home appliances, including schedulable and non-schedulable appliances. The main focus of the work in [25] was on the utilisation of building renewable energy in support of energy savings rather than fuel poverty.

The issue of fuel poverty is a multi-dimensional one which covers more than just the heating of homes. To provide a representative picture of how energy-awareness impacts older people, all aspects of energy consumption need to be considered. When looking at the problem of fuel poverty, any savings of an older person's energy expenditure must result in financial benefits. In turn, for a technological solution to fully address the issue of fuel poverty it must holistically look at energy-awareness as a whole, and provide ways in which behaviours and usage patterns can be altered to improve energy expenditure through an unobtrusive and proactive system which does not require constant user interaction to govern the control of the environment.

III. Design Considerations

Older people face a number of challenges in today's society which can impact their lives and well-being. Fuel poverty is a major societal challenge which affects significant proportions of the society. With limited financial resources available to older people who are reliant on state pensions, paying for fuel (i.e., electricity and gas) becomes a major concern for their well-being [4]. Within the UK alone, a high number of unnecessary deaths are caused each year due to weather concerns [2]. This can be due to excessively hot weather during the summer months or excessively low temperatures during the winter. In circumstances like this, which require a higher expenditure, it is common for the older person to go without basic heating or cooling due to limited financial resources, which enables them to focus on other costs (e.g., food and rent) [5]. However, this has a detrimental impact on their well-being due to the suffering involved, and possible death, during low or high temperature times during the year. Not only does it impact older people, but the costs to healthcare for dealing with emergencies relating to the admission and care of older patients at these times, places stress on services which are already over-stretched.

Communities and community care offer care services where carers or social workers visit the older person to ensure that they are coping and not suffering. However, with finite resources, only so much is able to be done. To help in the care of older people, smart technology has been employed quite extensively. For example, in [16], smart devices help monitor the older resident. However, relying on technology can introduce other challenges. More specifically, older people distrust and fear the use of technology [6]. This often leads to technology being ineffective due to the lack of engagement from the older person, especially if they are required to interact and engage with the technology in some way. Davis [6] introduces the Technology Acceptance Model which impacts on the perceived usefulness of technology by the older person which can introduce barriers for adoption. However, older users also encounter problems with a general lack of understanding of ICT which introduces barriers to the user as it impedes them from making strategic decisions when managing fuel costs [10]. This lack of understanding needs to be addressed for the system to be effective and allow seamless interaction with the user. For instance in [15][16] mobile technology is used but can introduce problems with engagement due to the fear of technology whereby the older person has to interact not only with an unknown and unfamiliar device but also some form of application and visual interface. If the interaction between the system and older person is perceived to be too difficult then a lack of engagement usually results [6][9]. Therefore, intelligent devices need to be unobtrusive as well as the system offering a less technological way of interacting with the older person to minimise the impact of the fear of technology.

Behavioural change is another key consideration for such systems as the technology would also have to monitor and determine if there are better behaviour patterns when interacting with energy-reliant devices within the home. For example, making a cup of tea or cooking dinner relies on some form of energy consumption (either electricity or gas) to complete the task. However, energy tariffs can change through the day and can impact on the total expenditure over a month. By modifying the behaviour of the older person to a more cost-conscious time can have benefits over the long term.

Therefore, when designing effective systems for reducing the energy consumption of older people, a number of design considerations have to be taken into account. Namely, the use of unobtrusive technology which is embedded within the environment; simple interaction devices which promote trust; behaviour modification and prediction to reduce or better manage the use of energy reliant devices within the home; and, systems to recommend changes to patterns.

This work proposes to address the question of whether an intelligent recommender system can reduce the energy consumption of older people by predicting and modifying their daily patterns and behaviours so that they minimise their energy needs to ensure an improvement in their general well-being by reducing anxiety associated with fuel poverty. Anxiety has a clear detrimental effect on the well-being of an older person [9], which will hinder their interactions and acceptance of technology. Improving the quality of life of older users is difficult to gauge as different people assign different values to things [8]. For that reason, the proposed system aims to reduce the fear of poverty and improve the quality of life for older users by providing them with guidance and options on how to change their behaviour so that they are more energy-conscious and aware of what they are doing through a simple house model interface. More detailed feedback can be given to those users who feel comfortable with technology or to carers, or family members who are actively helping the older user. By providing them with a more holistic and wide-ranging understanding of the energy-consumption which takes into account not only their heating but also their activities during the day can help them improve their overall energy consumption. By looking at the whole, more substantial savings can be made which in turn reduces their chances of falling into the fuel poverty trap.

An addition technical design consideration is due to the heterogeneous nature of the devices necessary to perform all the tasks necessary to monitor people in their home. No single commercial company yet provides lighting, power sockets, motion sensors and the other devices necessary for such a task. As such, it is necessary to create a technical solution that integrates the data from a range of available devices. This can be achieved in different ways, including runtime transcoding [20], brokers [21], architectures [18], cloud based solutions [19] and proxies. However, because of the relatively small scale of the required deployment, the data integration will be done in phase two which looks at data collection (see Section VII).

Therefore, in summary, this paper is intended to address the issue of the acceptance of technology as well as factors relating to the trust of intelligent systems and improving well-being by reducing fuel poverty. This will be through simple, trusted and unambiguous human-computer interfaces, behaviour and activity analysis; and, recommender systems to help promote more informed energy-aware decisions. This will allow older people to consider all aspects of their energy consumption, rather than focusing on one single aspect (e.g., heating) as their overall energy-footprint will be made of many different components. For example, heating, cooking, recreation, etc., which complements their potentially sedentary life-style. In addition, other system considerations when collecting telemetry from within the home on a more permanent basis will be looked at.

IV. Proposed System

The development of the system has been split into three distinctive phases. The first phase of development focused on the initial system development and interactions with the end user (see Section V). The second phase focused on providing more energy-aware devices for monitoring the usage patterns of older people (or test candidates) and system related issues. The final phase will be to deploy multiple sensors and interfaces into the houses of test candidates.

From a conceptual perspective, the system (see Fig. 1) will process sensor telemetry from devices located within the environment, determine if the behaviour of the person can be changed based on the task activity they are performing, which includes using tariff information gathered from energy suppliers through IoT devices and then facilitate in providing feedback to the older person.

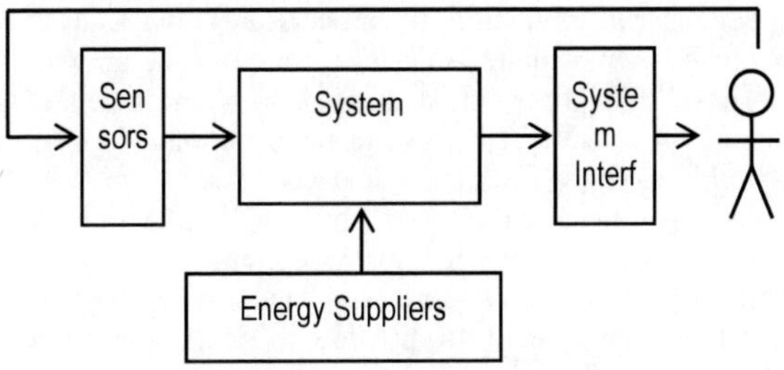

Figure 1. Overview of the proposed system.

The guiding principle in the design of the prototype is to determine if the energy consumption of the user can be reduced by determining the behaviour and patterns they are exhibiting. This builds on our existing energy-aware work which is looking into energy-aware programming languages and middlewares. The system relies upon a number of key components that deal with the interactions with energy suppliers to provide information, which informs the recommender system; activity/task detection and interfacing with the user through an unobtrusive interface. A system prototype has been developed which allows the seamless interaction with an older person, which could reduce their energy consumption. For this prototype iteration the full system was implemented but deployed within an idealised test environment to capture telemetry. The system was also able to save and playback telemetry for user evaluation. In the second iteration, data has been collected from a variety of energy-aware plugs, wireless enabled light bulbs and motion sensors. Information was collected over a seven-day period to give a richer data set of an occupant's activity, and how best behaviour and activity levels can be analysed for energy-aware decisions.

A. Sensors

A variety of off-the-shelf sensors have been used to provide contextual information regarding the state of the environment, the location of the older person, and what they are interacting with in the home environment. Sensors that provide telemetry on light, motion, pressure, electricity use, sound, as well as other sensors providing contextual state information of the environment, have been used. Sensors were typically subsumed within the appliances which consumed energy. This was mostly electrical appliances but scope for gas based appliances will be incorporated in a later phase. Sensors can be added seamlessly to the system by simply plugging them in and informing the software that a particular type of sensor has been added. Different sensors can be dynamically added, along with information which provides contextual information regarding how telemetry can be analysed and interpreted during the behavioural analysis phase. During the first phase, environmental and power usage sensors were based on Phidgets technology and were used to record and forward telemetry using an event system that had been developed (details in Section IV. B).

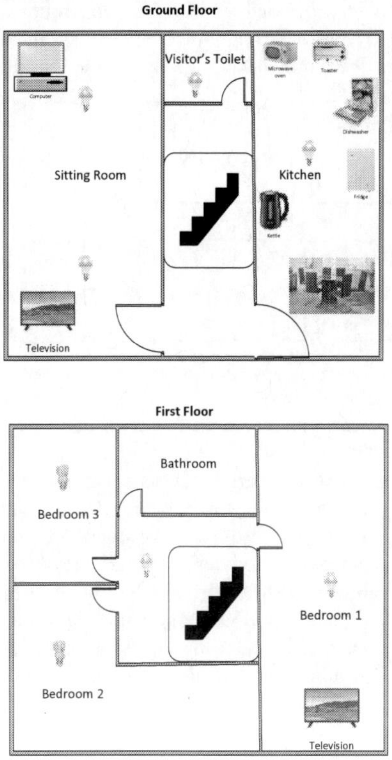

Figure 2. Deployment of wireless sensors.

Building on the first phase, a combination of smart Wi-Fi enabled energy plugs and other Wi-Fi enabled technology (motion sensors and wireless enabled light bulbs) were used. These are currently outfitted within a test house which is also being used for other work that is being currently investigating on energy-aware programming languages. Each room of the test house has a combination of motion, energy aware plugs and Wi-Fi enabled light bulbs. The present deployment consists of 8 energy aware plugs, 8 wireless light bulbs and three-in-one motion sensors for recording ambient temperature, light levels and motion environmental state information.

The test house has 7 rooms (including two toilets / bathroom) and a hall way spread over two floors. For the toilets / bathroom, only lighting levels were monitored. Sensors have been placed so that minimal occlusion occurs. Fig. 2 provides an outline of the deployment of wireless enabled sensors within the environment. There are a number of issues that will not be considering in this phase. This is principally dealing with privacy, occlusion and identify resolution issues.

B. *Behavioural Analysis*

To aid in the process of determining how much energy is being consumed within the house and whether a better alternative is available, telemetry from sensors are aggregated and processed within the behavioural analysis component of the system. An overview of the activity, behaviour and recommender stages is outlined in Fig. 3.

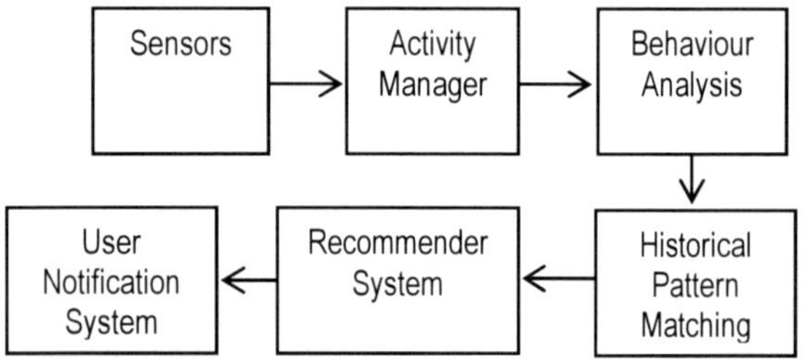

Figure 3. Activity, behavioural and recommender stages.

Activities are used to identify what task the user is performing within the home. For example, making a meal would involve using a variety of sensor equipped devices within the kitchen, each of which will be generating telemetry. In a typical use case, the oven, along with the cooker and supplementary devices (e.g., kettle) could be used. Combined with environmental state information originating from embedded sensors in the home, additionally telemetry regarding light, sound and motion information would be generated and be contextualised within the kitchen environment.

Activities are generated by end-users and allow the association between different telemetry and devices and a given activity. The approach that has been taken is to offer a simple means for users to define an activity and then to associate device combinations to that activity. These activities will then form the basis of the behavioural analysis and pattern-matching functionality whereby the system will be able to determine what type of activity is being performed by the user.

Raw sensor telemetry is tagged with additional meta-information regarding the device type, timestamp and location and is consumed by the system within an event message. For the first development phase, as the types of sensors were limited, events contained all contextual information which was needed to be consumed and processed correctly. During testing, this did not provide sufficient granularity for the types of devices used within the second phase of the system. Here, events became aggregates of the principle device types (e.g., plugs, lights and motion detectors, as outlined in the data structure behind the system in Fig. 4). Events are captured and processed within the behavioural component during the activity identification process. To aid in providing an environmental snapshot of what is happening within the environment, a window of applicability is used which allows events to be received over a short period of time. This populates activity states which are triggered when all associated sensor telemetry has been received during a set period. When this has been satisfied, activities trigger notifications, which in turn are processed by the analysis component. At this point, a decision can be made on what to do next. The first option is to process an associated action which causes the system to perform some form of real-world action which impacts the end-user. Alternatively, notifications can be used to build up composite events which provide more complex combinations of activities to be associated with each other. This allows the system to provide a richer and more complex activity detection process.

The system was enhanced in the second phase to capture more detailed information from each of the devices. This then interfaces with the rest of the system developed previously. The notable extensions in this phase are to allow for richer devices to be used. Namely, Hue and TP-Link based light bulbs which can provide contextual information regarding their intensity/brightness and, in the case of the Hue bulbs, different colours/shades. The latter can also be used to indicate mood information and will be incorporated in to the third phase of development and testing. Motion sensors have been used which expands on the coarse motion information which was available in the first phase of the system. Wireless based sensors which can record temperature, lighting and motion information are now used and provide contextual information which was spread over a variety of other sensors in the previous iteration. The system stores information (as outlined in Fig. 4) and links to the rest of the system developed in stage one.

Historical information regarding past activities is used to provide the ability to determine if activities have occurred during similar times in the past. For example, cooking food

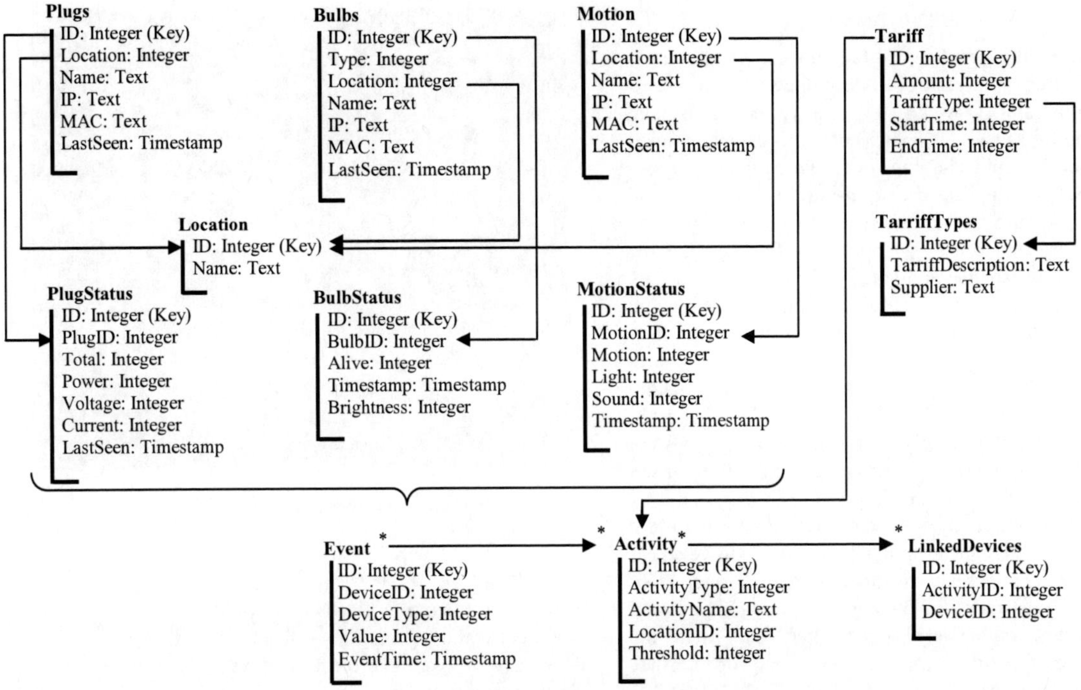

Figure 4. System data structure and relationships.

at specific times; use of kettle to make drinks; and activities that relate to their life-styles. This information is used by the recommender system to determine if user activity patterns can be modified to help improve the energy use within the home based on their occurrences. For instance, when coupled with tariff information, the recommender would be able to inform the user that by changing their pattern to do a particular activity by a few minutes could result in a tangible energy cost saving over a period of time. In the first phase of development, only small data sets were collected over a two day period, and then based on information generated on several devices. With the introduction of the second development cycle, devices were polled each second and data was stored. The number of devices was increased to incorporate the energy-aware plugs and light bulbs which were not used in the previous iteration. Because of this, it was found that the processing of historical information introduced additional issues (as discussed in Section VII). This was mainly due to the large size of data sets collected over the test period (now one week) where the method in which historical behaviours were identified and analysed would require more efficient storage and processing algorithms for them to work effectively in real-time.

C. Recommender System

The purpose of the recommender system is to provide end-users with feedback on their energy consumption performance or provide alternative actions. To best inform end-users of any energy cost savings, tariff information is used from the electricity supplier that they are signed up with. By analysing their past behavioural patterns, the recommender system will determine if any energy savings can be made by suggesting to the end-user how best to alter their habits. For instance, as a profile of the user is generated, any patterns or behaviours they exhibit will be identified and any improvements will be suggested. This information can be conveyed in a number of ways to the end-user. Notification messages detailing the current cost and potential savings can be shown to the user, whilst real-time feedback is given through a physical prototype system, which allows users to instantly see how much energy they are consuming within different rooms in the house.

An administrative component is provided which allows carers or family members to configure the system and provide guidance on what levels of energy expenditure the older user should use on average. This information is used to inform the older user that targets are not being met and to keep the carer or family member informed and involved in the care of the user. Additionally, these set targets also allow carers/family members to closely monitor that the older user is using a minimum threshold amount of energy, thereby ensuring that they are not suffering from the lack of heating or cooling.

V. User Interaction

Compounding the problem of fuel poverty, older users also suffer from technology acceptance. This adds additional constraints in which electronic systems subsumed within the home impact the experience and engagement of the older user. If the technology with which the older user has to interact is complicated or non-intuitive, this can lead to users disengaging with the system put there to improve their well-being and causes an opposite effect so that the user is even more isolated and stuck within the fuel poverty trap. The issue of reducing a lack of understanding and purpose of the system [10] needs to be considered to allow the prototype to be effective with users. In turn, Leonardi [7] highlights that the interactive medium needs to consider the "*motor and cognitive capabilities*" [7] of the older user. When considering the perceived usefulness of the system [6] to promote older users to engage and trust the system and recommendations, a clear simple interface is required. These issues are important when considering how the older person interacts with the prototype system. However, it does pose challenges when considering the acceptance of the interface as it must not rely on something that might cause confusion or anxiety when interacting with the system. Therefore, some form of mobile device or interface would pose significant challenges to older people.

Different interfaces were constructed and evaluated to determine which one offered the older user with the best way to interpret energy usage information. One method of conveying information was through a clock metaphor where LED's indicated the general energy usage within the home. However, this did not prove to be too useful or popular with older-users.

The approach, which ended up being taken, was to provide the older user with a simple, non-technical interface which does not rely on any interactive technology (e.g., mobile phones or smart televisions) but instead uses a simple traffic light metaphor. A mock-up miniature house was provided which uses a traffic light system located in each of the rooms within the model. Instant visual feedback is provided to the older end-user by indicating their power consumption in each room. Red indicates that they are excessively using energy within that room, amber indicates they are using more than they should but there is room for improvement with a few recommendations, while green indicates that they are using energy within the guidelines, or what the carer/family member might have specified.

Fig. 5 shows the model house prototype interface. Each room of the house contains a set of LED's which indicates the real-time use of energy within that room. This was found to be the most successful of the prototypes for conveying straight forward energy usage information to older users.

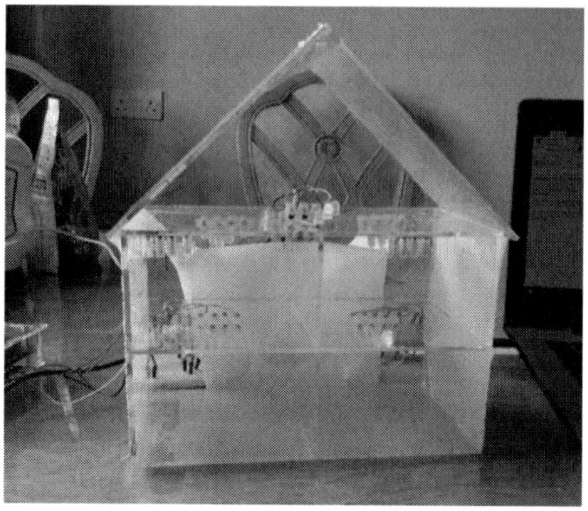

Figure 5. Model house prototype.

A. *Administrative interfaces*

The administrative interface provides carers or family members with cost saving's information either on a day-by-day basis or over a projected month. If older users felt comfortable with the technology they were also provided access to this information, rather than simply relying on the model house. For example, Fig. 6 shows the costs incurred for a number of monitored appliances within the home environment during the first phase of testing. Information is outlined, based on the activities that have been detected during the day and shows a comparison of their energy use. Both carers/family members and the more technically confident older user can utilise this information to determine how many times a day they perform specific activities with a view to reducing the frequency over time or to more suitable times which reduces their energy costs.

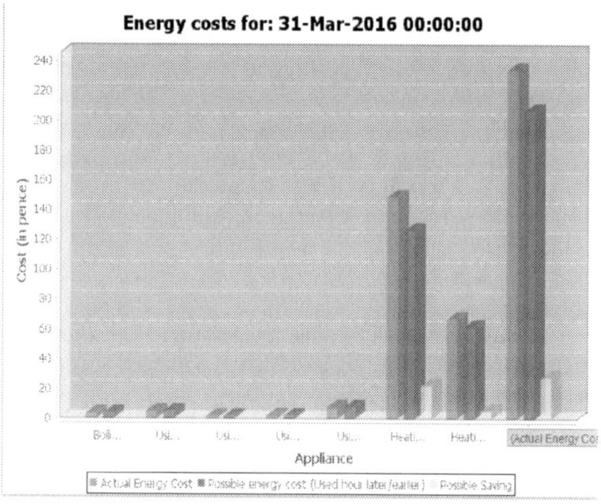

Figure 6. Savings graph (phase 1 of system).

Another way in which financial information can be conveyed to the user is through a monthly chart outlining the costs that have been incurred for each of the activities during that particular period. Fig. 7 shows how information can be presented to end-users regarding the total costs incurred and by activity during the first phase of testing.

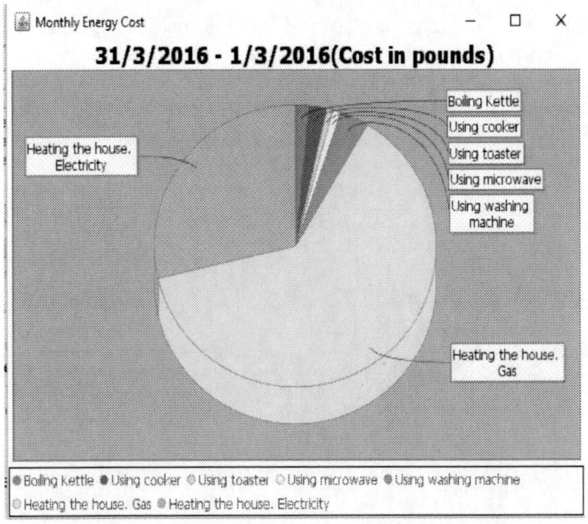

Figure 7. Monthly energy costs (phase 1 of system).

Recommendation notifications were also provided to end-users and carers/family members to allow them to determine where savings could be made by changing habits. For example, recommendations could be made to change the time of morning snacks and eating times to make best use of energy costs by moving these occurrences by a few minutes. For phase two, users were presented with different graph information representing their energy expenditure (examples of these are included in Section VII).

VI. Phase One Results

There are three principle phases to our testing of the system. The first phase was to collect basic data over a short period and test the effectiveness of the user interface (phase one). Phase two required spending a week within a test house monitoring energy, activity patterns and identifying system considerations e.g., storage and processing (phase two) while phase three will look at deploying within a number of homes for a longer period and expose users to the full monitoring system.

To combat fuel poverty, older users have to overcome issues with anxiety caused from the distress of lack of finance to pay for energy bills and also the fear of technology, which impedes the adoption of technology to help them become more energy efficient. Data was collected from an idealised test environment and played back through the system for evaluation purposes. The prototype was evaluated by 10 older users, whose ages ranged between 50 and 70 and were used to measure their technological fear and acceptance of using this technology. This was to determine whether monitoring energy use and recommending more efficient use of energy through modifying activity patterns, would improve their well-being by reducing their overall energy costs.

Preliminary testing was conducted using 10 users to determine the effectiveness of the model house prototype. Older users were asked to answer a questionnaire after evaluating the usefulness of the interface. The data collected represented activities spread over several days from the test environment and was played in compressed form for them to evaluate the prototype. Activities were highlighted and explained during the playback of telemetry to help identify what was being done. 100% of the respondents identified with what the model house prototype was attempting to do by making them more aware of their energy consumption patterns. It was found that 80% of users were able to interpret the real-time information with ease while 20% of the respondents had difficulty when the system detected an increase or decrease in the energy use and notified them through flashing LED's. When addressing the question of technology acceptance and the fear of using new technology which can beset an older user, 100% of respondents said that they did not feel anxiety or fear regarding accepting information generated through the model house prototype. When asked if the prototype provided them with a way to monitor and adjust their habits during the day to make better use of energy, all respondents replied favourably (100%). In fact, it was discovered at this point that a number of users would like even more information presented to them on improving their energy expenditure. This proved to be encouraging and a validation that the model house improved their energy awareness and reduced their anxiety from worrying about fuel costs. Another group of 10 users (aged between 20 and 50) were used to evaluate the usability and functionality of the backend functionality of the system. This type of functionality would be used by carers or family members to help advise the older end-user on how to improve their energy-use by exposing them to fine-grained data regarding which activities were done and how potential savings could be made. The types of questions asked to these participants were predominantly focused on activity management, savings information, monthly projections and recommendations. The results from this evaluation showed that 100% of participants understood what the activities were and how they related to the system and data while 70% of participants were able to deal with, and manage, activity related features. The remaining 30% required extra guidance before they felt fully comfortable with the back-end system. When addressing the usefulness of the recommendations and potential financial savings, all participants agreed that the system was easy to use and offered important and helpful information on how to improve energy usage.

VII. Phase Two Results

Our second phase of development and testing was spread over a seven-day period and collected data from a test house (see Fig. 2). This phase primarily focused on the integration of heterogeneous IoT devices, collection, storage, processing and administrative analysis graphs. An expansion on the number the monitoring capabilities of the initial system was undertaken by incorporating off the shelf components that are readily available instead of relying on specialist hardware. More devices would be monitored, as well as lighting patterns and other information. This phase also considers the additional system based issues that were not encountered in the first stage due to the limited data collection and primary focus of interacting with the older user.

Electrical appliances were plugged into energy monitoring wireless plugs, while standard light bulbs were replaced within the house environment with equivalent wireless light bulbs. Complementing the environment, each room had a wireless motion sensor (which recorded motion, lighting and temperature telemetry) in the main living areas (front/living room, kitchen and bedroom). Telemetry was captured over a 24-hour, 7-day period. Each wireless device was given unique identifiers so that energy use could be tracked for each associated electrical appliance as well as the occupancy of the room. The house is occupied by two people (male and female) and ages range from 40's to 60's. The selection of the appliances to monitor was based on the availability of energy-aware plugs. The rooms with high energy use and occupancy were identified as the kitchen, front room and bedroom. Therefore, the appliances that were chosen were located within these rooms. The devices monitored in the kitchen were: microwave, kettle, toaster, fridge/freezer and dish washer. Ideally, the cooker and washing machine would have been monitored but these were found to be wired directly into outlets rather than being plug based. These will be monitored when running the third phase in three separate homes. In the front room, the devices which were monitored were the TV and computer. In the bedroom, a TV was monitored.

The purpose of this stage was to collect telemetry which can then be used to test system and data integration more fully before monitoring the usage of a number of older people. As telemetry was collected during the summer months, devices which aided in cooling the occupants down (i.e., fans) would have been ideal. However, the test home did not contain these. At colder times of the year, other devices that provide heat (e.g., electric blankets, central heating) would be considered and will be monitored during phase three. Motion sensing was limited to the main areas of the home (front room, kitchen, and main bedroom) and provided contextual information regarding light and temperature levels.

The results for each monitored room are shown in Fig. 8 for the Kitchen, Fig. 9 for the main living room and Fig. 10 for the bedroom. It was observed that an issue occurred during day 5. This was down to the Wi-Fi router becoming unavailable and was spotted by one of the occupants who reset it. The subsequent figures highlight the loss of telemetry during this time.

Fig. 8 shows the energy usage of the monitored devices located within the kitchen for a 7-day period. The monitored devices consist of a microwave, kettle, a fridge/freezer, toaster and dishwasher.

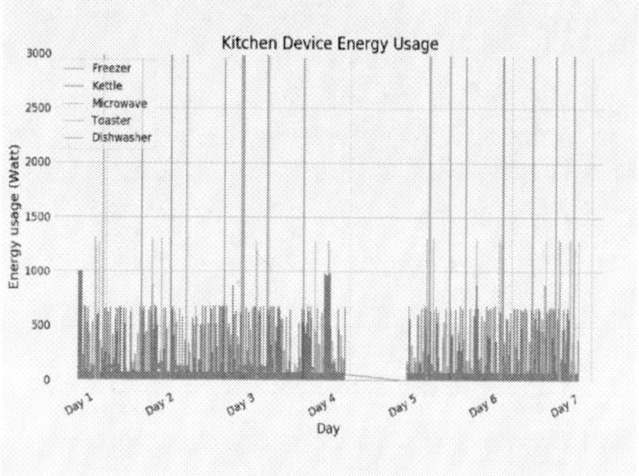

Figure 8. Kitchen device energy use.

The usage patterns and properties of each device are correctly identified for the room. The freezer has a constant load of around 50 Watts, with regular peak loads of around 600 Watts – as expected when the cooling circuits are in operation. The microwave is generally not used heavily, mainly just for preparing evening meals, so only appears at its peak load of 1200 Watts. The kettle is likewise not used regularly, appearing when used at a load of around 3000 Watts for short amounts of time. The toaster is used intermittently during the day and usually at the same time as the microwave. These times are around the time that food is prepared in the morning and afternoon/evening. The peak load of the toaster is 1000 Watts. The dishwasher is used once or twice a day and peak load is up to 250 Watts. Based on the results, and the frequency of the more commonly used items (i.e., kettle, and microwave), the system would be able to suggest times at which energy costs are lower and thereby, reduce overall costs by time-shifting to more affordable time periods. The same applies to the frequency of the dishwasher activity, where changing the times of day can result in cost savings. On the other hand, the fridge freezer which is on continuously due to the nature of the appliance would not be something that can be time-shifted. However, information would be available to provide whether the appliance is indeed energy-efficient to help identify cost savings if a newer, energy-efficient appliance was used instead.

Fig. 9 shows the energy use of the monitored devices located within the front room for a 7-day period. The monitored devices consist of a TV and a computer. The TV tends to be used briefly in the morning and lunch, and more extensively in the evening. It has a relatively high standby of around 15 Watts, and an on-power usage of around 150 Watts. The computer is mainly switched on in the evening. It has a very low standby power usage and an on-power usage of between 100 and 180 Watts.

As only two devices were monitored in the living room, the most commonly used electrical appliances in use were the TV and computer. It can be seen from fig. 9 that the TV is in use at different times of the day where it is mainly used towards the morning and early afternoon. The associated energy costs for this would be higher due to the higher energy costs from the supplier, so changing habits could result in savings. In the afternoon and evenings, both the computer and TV were in operation, although on some days, their usage was spread more evening across the day rather than clustering exclusively in the morning / afternoon / evening areas. Changing the habits of the user could result in cost savings when considering that both devices are used extensively during the day, and at times, concurrently.

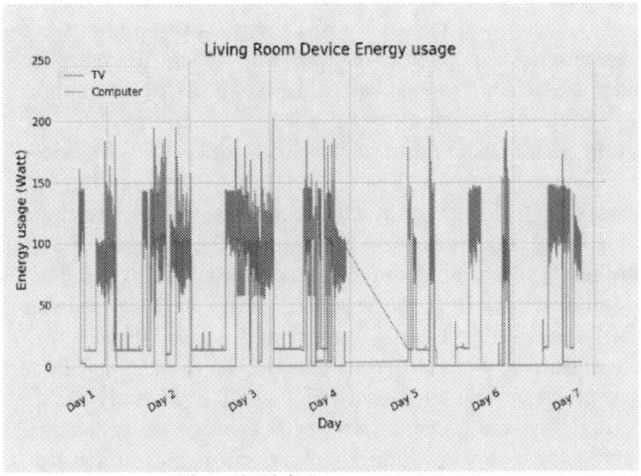

Figure 9. Living room energy use.

To enable further analysis of energy usage, the cumulative power usage of these two rooms can be plotted based on the collected data. Fig. 10 shows the cumulative energy usage in Watt-hours (Wh) of the devices in the kitchen. It shows the cumulative power usage from the devices being generally steady. This information would provide the user with an idea of the total energy consumption for the duration through the administrative interface. The accumulated energy use of the toaster shows that there was an issue with the plug storing the daily energy totals. It does match with the activity patterns in Fig. 8 where the toaster was used on day 1 and day 4 with no activity between those days.

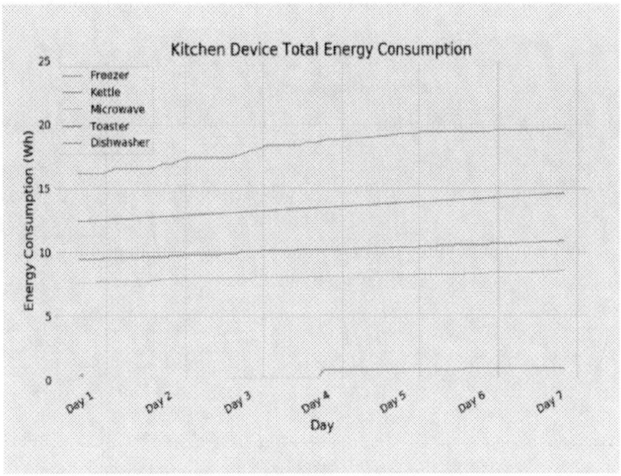

Figure 10. Cumulative kitchen device energy usage.

Fig. 11 shows the cumulative energy usage in Watt-hours (Wh) of the devices in the living room. It shows a steady increase in power usage.

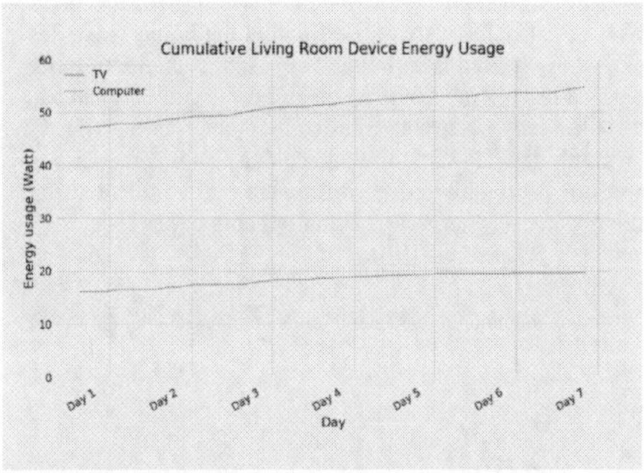

Figure 11. Cumulative living room device energy usage.

Fig. 12 shows the cumulative energy usage in Watt-hours (Wh) of the device(s) in the bedroom. Although only one device in the bedroom was monitored, the usage activity of the TV indicates rough times at which the occupants are occupying the room. This also can provide loose information about the times at which they sleep (before and after activity through TV usage). The peak load of the TV is around 80 Watts when in use for short periods of time.

Utilising more devices located within the bedroom would provide more contextual information regarding the habits of the occupants.

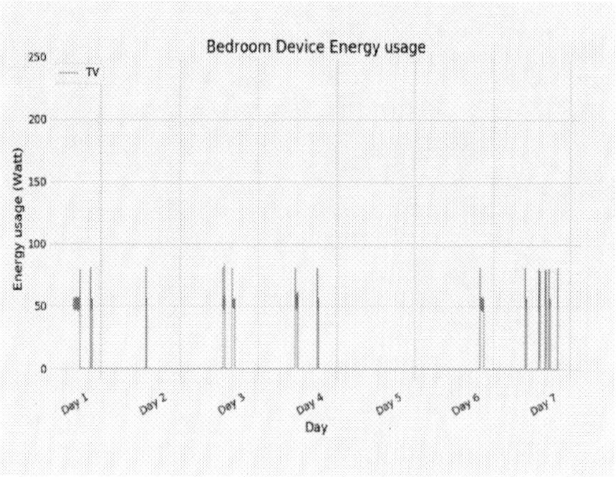

Figure 12. Bedroom energy use.

As well as monitoring the energy usage from electrical appliances, the lighting in the house was also monitored. The use of lights during the test period is shown in Fig. 13. Two types of bulbs were utilised within the home. Five TP-Link Wi-Fi enabled light bulbs were used in the main living areas. Three Hue bulbs were used in addition in the bathroom, and as supplements to the other rooms. Fig. 13 shows only the TP-Link bulb usage during the test phase as the information gathered from the Hue bulbs indicated that while not in use, energy was actually being expended.

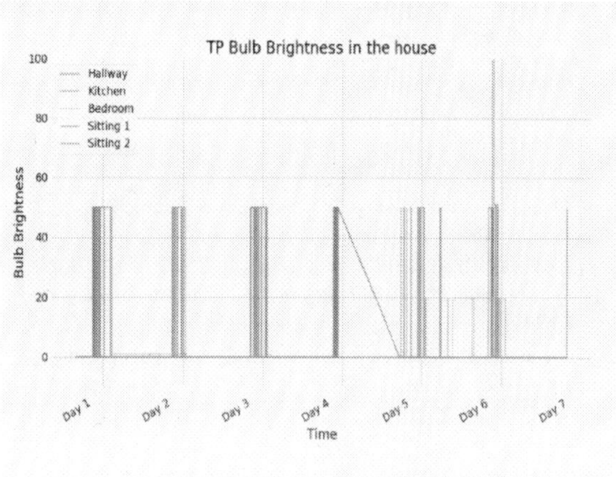

Figure 13. Light usage across the house.

In Fig. 13, it was decided to layer all the TP-Link lights in the household on the same figure, as it illustrates an important point. The household lighting is almost always switched off. There is a narrow window of 2-3 hours a day when the lights are activated which correlates with the longer days during the summer months when the data was collected. This would be greater during the winter months due to the shorter daylight hours and would suggest even more power usage efficiency. The diagonal lines represent data collection issues due to the bulbs dropping the Wi-Fi and presenting erroneous results. It is believed that the most efficient approach to reducing lighting waste is to install cheap motion sensors that activate, and de-active lights based on movement and ambient lighting. It is worth noting that the lighting data was also collected over summer, hence their minimal usage.

During this phase of testing, other issues outside of energy-awareness have been shown to be important and need to be addressed further. Issues concerning the size of the data set generated from a variety of devices and lights resulted in a database exceeding 124MB in size for the one week of testing (8 devices & 8 lights). Based on this, for a system running realistically over a year, the amount of data generated would be in the range of 6448MB. With more devices being added which are polled for telemetry, this would increase the amount of information generated. This would mean that efficiency and storage issues would come into play which was not considered in the first phase of development. The time to process information to present to end users via the administration interfaces would take considerable time and as a result, would need to implement more efficient ways on processing vast amounts of information. The same can be said when it comes to analysing behaviours and activities over a period of time for the system to provide useful recommendations. As the second phase was within a closed system (i.e., all devices were connected to a Wi-Fi router with limited internet access), any information requiring being stored on the cloud and its subsequent processing would become an issue. Another Wi-Fi related issue observed from the test data is the intermittent nature of the router itself. The one used was a fairly cheap off the shelf device but did show that the Wi-Fi signal dropped at different times. This would introduce the problem of lost telemetry which could potentially have a bearing on the recognition and analysis components. Making the collection and aggregation of this information from devices more resilient and able to cope with periods of no connectivity would need to be addressed. However, it is envisaged that the system will store and process information locally within the house. Issues regarding the privacy ~~of~~ and security of this information also pose an interesting dilemma. The information collected provides a very fine snapshot on the behaviours and activities of the occupants, and thus, privacy and the safeguarding of this information is paramount. These issues will be considered in phase three of the system development.

VIII. PHASE THREE RESULTS

The final phase of testing will be conducted over the next several months with live participants. A small number of candidates will be selected which will allow us to analyse the effectiveness of the system. Both male and female test subjects will be selected so that analysis of the differences between genders, age, disability (both physical and cognitive related) and living status can be considered. In addition, system based issues regarding storage, processing, resiliency and privacy will be explored to best protect the end user from potential abuse of information. The intention is to outfit three homes with larger numbers of energy-aware plugs, motion detectors and wireless light bulbs over at least a two-month period.

IX. DISCUSSION

The goal of this work is to look at energy-awareness and how this impacts older users. Our intention was to produce a simple ICT system which would attempt to reduce the anxiety that older people experience in relation to fuel poverty as well as address the issue of accepting technology. It was found that the prototype system promoted the perceived usefulness of the system to older users. This is in addition to other areas of our research into energy-aware programming languages and runtime systems.

By focusing on the energy-awareness of everyday use from the start, the prototype system was able to successfully improve the energy use of a user over the initial trial period. The behavioural analysis and recommender systems were found to accurately identify activities which in turn aided in recognising and highlighting the cost awareness of what users were doing in association to their activities. By increasing this cost awareness of their actions, users were found to be more aware of how to reduce their energy costs by changing times or common activities to more cost-effective times of the day. Carers/family members would be able to help by raising awareness of what the older user was doing on a day-to-day basis by showing them how simple changes to their daily patterns could result in tangible cost savings as well as improved health.

In the second iteration, the primary focus was on collecting richer data sets with off the shelf sensor technologies that are commonly available. The devices that were used during this process were TP-Link energy aware plugs, TP-Link bulbs, Hue bulbs and motion sensors. A test house was outfitted with a number of rooms containing these smart devices and data was collected over a 7-day period. The energy use of the test subjects in the principle living areas (kitchen, front room and bedroom) was monitored. Although only 8 electrical appliances were monitored, usage showed basic patterns and activities during the day and night. Light usage was also monitored but as the experiment was performed during the summer months, limited lighting patterns were shown. The amount of data collected over one week showed that additional investigation is required in the storage and processing of large data sets to provide a responsive system, as well as the privacy and safe guarding of intrusive information and resiliency of transmission / reception of information.

The next phase is to introduce more technology into three homes and monitor more electrical appliances. This is currently underway, and with the intention to capture several months of usage data to validate the system and house interface. Experiments on how the system affects the overall well-being of the older person will also be conducted as well as the impact of gender, disability (both physical and cognitive related), living status and age on their acceptance and use of the technology.

X. CONCLUSION

This paper has argued the case for simpler interaction and energy-aware systems for helping older members of the community to be aware of, and monitor their energy expenditure. This is a relevant societal problem which is addressing the anxieties of fuel poverty facing vulnerable groups of individuals as well as their acceptance of technology. An initial prototype system has been introduced as well as the lessons learnt from this first phase along with a subsequent data collection phase detailing the testing and analysis of a bigger data set. Following this, the considerations for the future direction of the work has been outlined. In summary, the benefit of energy-aware systems has shown to be a positive influence of the well-being of an older person by reducing their anxiety concerning interacting with technology and reducing costs associated with fuel poverty.

REFERENCES

[1] J. Robinson, K. Appiah, and R. Yousaf, "Improving the well-being of older people by reducing their energy consumption through energy-aware systems," in Proceedings of the 9th International Conference on eHealth, Telemedicine, and Social Medicine (eTELEMED 2017), Nice, France, March 2017. *ISBN 9781612085401*

[2] Department of Energy & Climate Change, "Fuel Poverty Report 2015," Available at: https://www.gov.uk/government/uploads/system/uploads/attachment_data/file/468011/Fuel_Poverty_Report_2015.pdf [retrieved: November, 2016]

[3] G. Flood, (2013). "British government's 'Digital By Default' plans challenged," Information Weekly (Online), available at: http://search.proquest.com/docview/1319367398?accountid=14693 [retrieved: November, 2016]

[4] The Poverty site, "UK: excess winter deaths," The Poverty Site. Available at: http://www.poverty.org.uk /67/index.shtml [retrieved: November, 2016]

[5] Greenseniors.eu, "Green Seniors In The European Union | Ageing Is Nothing For Cowards (May West)," Available at: http://www.greenseniors.eu [retrieved: November, 2016]

[6] F. Davis, "Perceived Usefulness, Perceived Ease of Use, and User Acceptance of Information Technology," MIS Quarterly, vol. 13, no. 3, pp. 319, 1989.

[7] C. Leonardi, C. Mennecozzi, E. Not, F. Pianesi, and M. Zancanaro, "Designing a familiar technology for elderly people," Gerontechnology, vol. 7, no. 2, 2008.

[8] M. Farquhar, "Elderly people's definitions of quality of life," Social Science & Medicine, vol. 41, no. 10, pp. 1439-1446, 1995.

[9] E. De Beurs, A. Beekman, A. Van Balkom, D. Deeg, and W. Van Tilburg, "Consequences Of Anxiety In Older Persons: Its Effect On Disability, Well-Being And Use Of Health Services," Psychological Medicine, vol. 29, no. 3, pp. 583-593, 1999.

[10] R. Ellis and J. Allaire, "Modeling Computer Interest in Older Adults: The Role of Age, Education," Computer Knowledge, and Computer Anxiety. Human factors, vol. 41, no. 3, pp. 345-355, 1999.

[11] Hivehome.com. "Hive," Available at: https://www.hive home .com, [retrieved: November, 2016]

[12] IFTTT.com, IFTTT", Available at: https://ifttt.com/ [retrieved: November, 2016]

[13] Honeywelluk.com, "evohome - Honeywell UK Heating Controls," Available at: http://www.honeywelluk.com/products /Underfloor-Heating/evohome-Main/ [retrieved: November, 2016]

[14] Heat Genius, "Heat Genius remote heating control benefits: save money, save energy, save time," Available at: https://www.heatgenius.co.uk/benefits/ [retrieved: November, 2016]

[15] J. Kim, H. Choi, H. Wang, N. Agoulmine, J. Deerv, and W. Hong, "POSTECH's U-Health Smart Home for elderly monitoring and support," IEEE International Symposium on a World of Wireless Mobile and Multimedia Networks (WoWMoM), pp. 1-6, 2010.

[16] T. Hargreaves, M. Nye, and J. Burgess, "Making energy visible: A qualitative field study of how householders interact with feedback from smart energy monitors," Energy Policy, Volume 38, Issue 10, pp. 6111-6119, ISSN 0301-4215, October 2010.

[17] A. Dohr, R. Modre-Opsrian, M. Drobics, D. Hayn, and G. Schreier, "The Internet of Things for Ambient Assisted Living," 2010 Seventh International Conference on Information Technology: New Generations, Las Vegas, NV, 2010, pp. 804-809. doi: 10.1109/ITNG.2010.104

[18] W. Wang, K. Lee, and D. Murray. Springer, "A Global generic architecture for the future Internet of Things," Service Oriented Computing and Applications, SOCA (2017). https://doi.org/10.1007/s11761-017-0213-1

[19] K. Lee, D. Murray, D. Hughes, and W. Joosen, "Extending sensor networks into the cloud using amazon web services," Networked Embedded Systems for Enterprise Applications (NESEA), 2010 IEEE International Conference on. IEEE, 2010.

[20] K. Lee and K. Gilleade, "Towards Generic Data Processing of Real-time Physiological Data in the Cloud," International Journal of Big Data Intelligence, Special Issue on: ``E-Health Systems and Semantic Web" Vol. 3 No. 4, pp 215-227, Inderscience, October 2016, DOI: 10.1504/IJBDI.2016.10000807

[21] W. Wang, K. Lee, and D. Murray, "Integrating Sensors with the Cloud using Dynamic Proxies," 23rd Annual IEEE International Symposium on Personal, Indoor and Mobile Radio Communications (PIMRC 2012), Sydney Australia, September 2012, DOI: 10.1109/PIMRC.2012.6362579.

[22] Y. Thongkhao and W. Pora "A low-cost Wi-Fi smart plug with on-off and Energy Metering functions," 2016 13th International Conference on Electrical Engineering/Electronics, Computer, Telecommunications and Information Technology, 2016, pp. 1-5. doi: 10.1109/ECTICon.2016.7561264

[23] J. Granados, A. M. Rahmani, P. Nikander, P. Liljeberg, and H. Tenhunen, "Towards energy-efficient HealthCare: An Internet-of-Things architecture using intelligent gateways," 4th International Conference on Wireless Mobile Communication and Healthcare - Transforming Healthcare Through Innovations in Mobile and Wireless Technologies, 2014, pp. 279-282. doi: 10.1109/MOBIHEALTH.2014.7015965

[24] A. M. Rahmani, T. Nguyen Gia, B. Negash, A. Anzanpour, I. Azimi, M. Jiang, and P. Liljeberg, "Exploiting smart e-Health gateways at the edge of healthcare Internet-of-Things: A fog computing approach," Future Generation Computer Systems, 2017, ISSN 0167-739X, http://dx.doi.org/10.1016/j.future.2017.02.014.

[25] B. Zhou, W. Li, K. Wing Chan, Y. Cao, Y. Kuang, X. Liu, and X. Wang, "Smart home energy management systems: Concept, configurations, and scheduling strategies," Renewable and Sustainable Energy Reviews, Volume 61, 2016, Pages 30-40, ISSN 1364-0321, http://dx.doi.org/10.1016/j.rser.2016.03.047

[26] L. Mano, B. Faiçal, L. Nakamura, P. Gomes, G. Libralon, R. Meneguete, G. Filho, G. Giancristofaro, G. Pessin, B. Krishnamachari, and J. Ueyama, "Exploiting IoT technologies for enhancing Health Smart Homes through patient identification and emotion recognition," Computer Communications, Volume 89, 2016, Pages 178-190, ISSN 0140-3664, http://dx.doi.org/10.1016/j.comcom.2016.03.010

Blended Learning Improves Physical Assessment by Nursing Students

Machiko Saeki Yagi
School of Nursing
Jichi Medical University
Shimotsuke, Japan
email: mati-s@umin.ac.jp

Shigeki Tsuzuku
Research Center for Instructional Systems
Kumamoto University
Kumamoto, Japan
email: tsuzuku@kumamoto-u.ac.jp

Natsuko Miura
School of Nursing
Iwate Prefectural University
Takizawa, Japan
email: natsuko@iwate-pu.ac.jp

Alan Kawarai Lefor
School of Medicine
Jichi Medical University
Shimotsuke, Japan
email: alefor@jichi.ac.jp

Abstract—**The ability to perform accurate physical assessment is an essential skill, which also contributes to developing the nursing process, and providing appropriate care. However, patient safety initiatives in some institutions limit opportunities for nursing students to receive adequate training. A blended learning system combining e-learning with simulation may provide the necessary training. However, learning outcomes for these methods have not yet been reported. We undertook this study to explore learning outcomes using a blended system to improve physical assessment skills of nursing students in Japan. The contribution of e-learning to improved physical assessment skills was evaluated in depth, as well as outcomes after training and the influence of blended learning on the desire to learn. From 82 junior nursing students who had completed physical assessment training, we recruited 46 students for this study. Each was assigned to one of two groups: a blended learning group or a simulation-only group. Objective assessment and subjective evaluation of the e-learning experience was performed during the study including a checklist by a proctor, and assessment at skill stations. All participants were able to greet the patient upon entering room and exit the room after a polite goodbye. All participants in the blended learning group were able to measure the patients' vital signs and interpret the results. Scores of those in the simulation-only testing the ability to perform skills for assessment of side-effects of chemotherapy and excretion condition were lower than the blended group (88% vs 17% and 74% vs 17%). Participants in the blended group were more likely to make recommendations around "lack of balanced nutrition" and "risk for infection", as "consultation can help ensure that a diet plan is enacted", "Proposal to have an antiemetic administered", " Initiate hand washing", "Wearing mask", and " Initiate mouth washing ", while those in the simulation-only group were more likely to make recommendations regarding "Anxiety" and "follow-up examination". These results suggest that e-learning improves the ability of nursing students to evaluate vital signs and information gathering skills. Simulation training was effective because e-learning stimulated participants' imagination of the simulation content. However, e-learning itself had no significant learning effect and simulation training enabled participants to perform the actual skills. Blended learning helped participants recognize the relevance of learning content to their professional practice. E-learning is effective in the study of assessment skills but must be combined with simulation to improve actual physical assessment, as a blended learning program.**

Keywords- nursing students; physical assessment; nurse education.

I. INTRODUCTION

Nurses are increasingly expected to advance their skills of patient assessment, particularly the ability to provide advanced and accurate physical assessment [1][2]. This change is prompted in part by changes in the healthcare environment, such as advancements in medical technology, increased age and disease severity of patients, shortening of the average length of hospital stay, and an increasing transition to home care. Thus, nurses must begin physical assessment training that is suitably adapted to clinical practice (e.g., simulation education using simulators, and simulated patients) as early as possible, ideally in their basic nursing education. However, simulation and clinical practice require more practice materials, more instructors, and extra time which increases instructor hours per student [3]. As physical assessment relies on nurses appropriately integrating knowledge and technique, such as conducting an accurate physical examination and judging a situation based on available information, learning to perform these skills is challenging [4]. Optimal methods for teaching integrative content to large groups of people need to be examined to improve physical assessment education.

E-learning may be effective to ensure optimal learning for each individual in group learning situations. In the context of continuing education for physicians, e-learning has been demonstrated to be as effective as conventional teacher-led education [5].

E-learning appears to be an effective method of instruction, which is not limited by the learning environment or time for learning, at least for knowledge-level content. We hypothesized that e-learning would be appropriate for teaching physical assessment skills, and in particular the knowledge-level component of such skills, to nurses.

In recent years, the utilization of e-learning in medical education has been expanding, marking a shift from passive knowledge transmission to active, self-motivated (i.e., lifelong) learning [6]. Self-motivated learning using e-learning tools has become a critical feature of nursing education. Learning which effectively utilizes information and communication technology needs to be implemented at the early stages of nursing education. Students receiving basic nursing education are considered "digital natives". They are from a generation for whom the internet has existed since birth [7], indicating that their information literacy is higher than that of past generations. They have a high likelihood of readily embracing e-learning in their education.

Although e-learning affords numerous benefits to learners, it also presents a number of challenges. First, e-learning relies on a learners' ability to self-manage the learning process, and second, e-learning methods must possess features to maintain the desire to learn [8]. For example, massive open online courses (MOOCs) [9] can theoretically provide instruction to an extremely large number of participants, but they do not encourage learners to self-manage learning, and students tend to drop out of these programs due to diminished motivation. A strategy proposed to resolve these problems in MOOCs is "blended learning," which combines face-to-face and e-learning methods. We hypothesized that a blended learning approach may help overcome the challenges of e-learning. This aim of this study is to investigate the effects and influence on motivation of a blended learning system to improve physical assessment skills by nursing students. The remainder of this paper is organized as follows. Section II presents the methods. Section III provides an overview of the results. Section IV discusses implications and limitations. Finally, conclusions and future work are detailed in Section V.

II. Methods

A. Participants and Data Collection

From 82 junior nursing students who had already completed physical assessment training, we recruited 46 students who consented to participate (45 females and 1 male aged 20 to 22 years). Each participant was assigned to one of two groups: a blended learning group or a simulation-only group (Figure 1). Preliminary evaluation of participants' computer literacy revealed that they were able to connect to the Internet and could effectively use mobile devices (e.g., laptop, smartphones), but had no previous e-learning experience. Additionally, they either did not have any experience with simulation training or had undergone it only once or twice. Data collection was conducted through objective assessment (i.e., correct answer rate on an e-learning pre-test, a checklist evaluating simulation results, correct answer rate on an e-learning post-test) and subjective evaluation of the e-learning experience (questionnaire and interview). Training was conducted from September to October 2013.

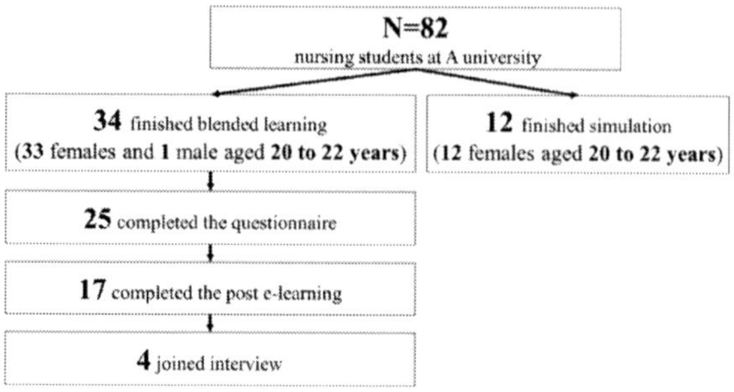

Figure 1. Study Flow

B. Design of Physical Assessment Training Using E-learning and Simulation

Learning Objectives: The objective of physical assessment training was to be "able to visit the patient's bedside and assess their condition." We created a checklist of skills that participants should be able to perform at the end of the training, including visual inspection, performing a medical interview, measuring vital signs, and performing auscultation and palpation. We created a simulation scenario wherein they conducted physical assessment of a patient with breast cancer receiving chemotherapy. This blended learning scenario was designed to create suitable educational materials to help nursing students understand real-world clinical situations, by using stories to help them better anticipate the realities of clinical practice. The characteristic points of the materials were the use of Japan's national nursing examination, modified using scenarios and actual clinical cases, and story-type e-learning that included learning opportunities for assessment, clinical reasoning, patient safety, and communications.

E-learning Design: The open-source learning management system, Moodle [8], was used to design the e-learning system (Figure 2). To eliminate differences according to usage environment, the e-learning system presented learning materials in a text format and images using Japanese manga comics only (i.e., no video) so that the system could run smoothly on PCs as well as mobile devices such as smartphones and tablets. The purpose of using manga in this study is to help participants imagine the patient's condition and the situation in a simulated hospital room. Participants were tested first to evaluate their baseline ability. The e-learning system presented a simulated patient, and the participant was expected to perform a physical assessment using the displayed data. The system also included a test on content related to the medical interview, physical assessment, and communication about findings, interventions, and patients' status to a supervising nurse. The test was conducted before simulation training to confirm if they possessed the knowledge necessary to take part in the simulation. As in previous research [11], the structure of the e-learning system involved solving problems in a format similar to the simulation, thus serving as an introduction to the simulation environment. The e-learning covered the entire nurse-patient interaction including entry to the patient's room, conduct of physical assessment, and exit from the room. It was configured such that participants answered 14 multiple choice questions and two short-answer questions, for a total of 16 questions (Figure 3 and Table I). Patients' vital signs in the e-learning post-test were different from the pre-test to discourage learning vital signs by rote.

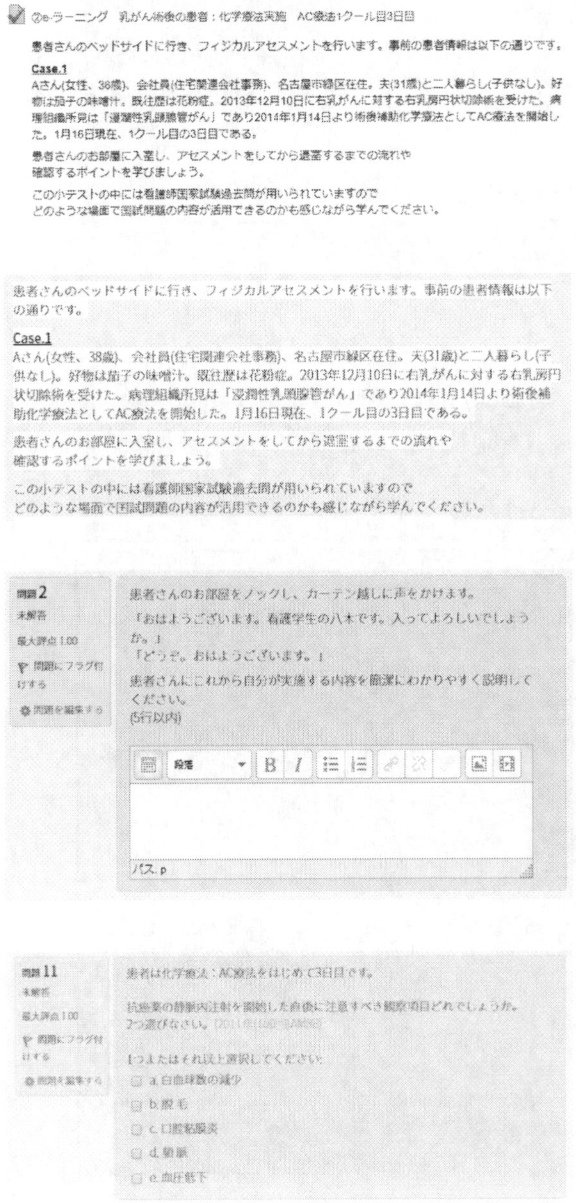

Figure 2. (a,b,c,d) e-learning interface (Japanese)

TABLE I. E-LEARNING TEST CONTENTS

Number	Contents	Methods
1	The rule of entering a hospital room	Multiple choice question
2	Explaination of implementation	Free coments
3	Correct site of measurment of body temperature	Multiple choice question
4	Nomal range of body temperature	Multiple choice question
5	Correct site of measurment of pulse	Multiple choice question
6	Nomal range of pluse rate	Multiple choice question
7	Correct position for post-ope patient about breast cancer	Multiple choice question
8	Measurement methods of blood pressure	Multiple choice question
9	Nomal range of blood pressure	Multiple choice question
10	Nomal range of respiratory rate	Multiple choice question
11	Side effect of chemotherapy	Multiple choice question
12	Communication with patient feels vomiting	Multiple choice question
13	Check point for patient during chemotherapy	Multiple choice question
14	Communication with constipation patient	Multiple choice question
15	Physical examination about abdomen	Multiple choice question
16	Explaination about physical findngs and diagnostics releted to patient condition	Free coments

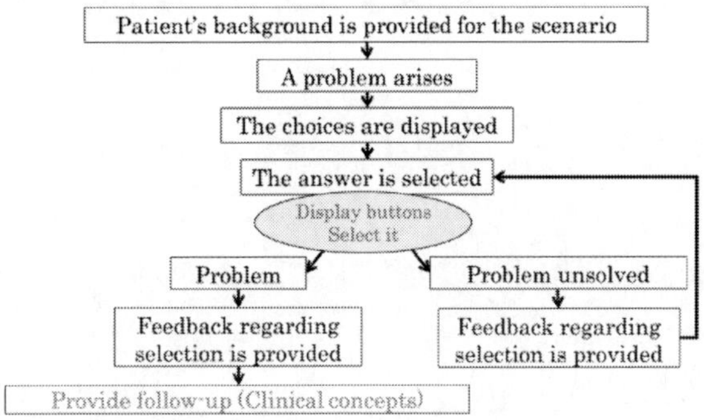

Figure 3. E-learning Flow

Simulation Design: The simulation was conducted in a simulated hospital room (Figure 4). An advanced simulator that enables the measurement of patients' blood pressure, pulse, respiratory rate, pupil size as well as auscultation of lung and bowel sounds was used (ALS Simulator, Laerdal Medical Japan Co., Ltd.). The simulator included skin moulage consistent with the appearance of a patient who underwent breast-conserving surgery. In the medical interview, the patient's verbal responses were produced by an instructor with a microphone and the voice came from a loudspeaker built into the simulator in order to impart a feeling of realism, as if the subject was speaking with the simulator directly. The procedure for the simulation was explained beforehand and an orientation was conducted, wherein subjects touched the simulator, measured its blood pressure, etc. in the simulated hospital room to eliminate unnecessary stress due to lack of familiarity with the simulation process [12]. Simulation training was conducted for 10 minutes per person. Participants then used the simulation training to assess the patients' condition and reported using the SBAR (situation, background, assessment, and recommendation) method to the instructor. [13]. During the simulation, performance was evaluated using the simulation checklist. Feedback was given to participants and participants took an e-learning post-test to assess their knowledge after the program (Figure 5).

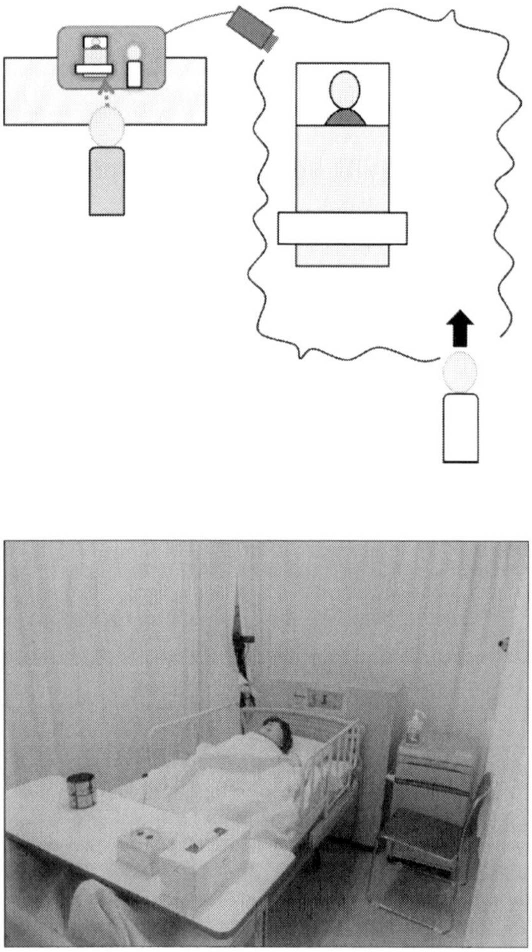

Figure 4. (a) The instructor watches participants' performance via video to ensure a safe learning environment (b)Simulated hospital room

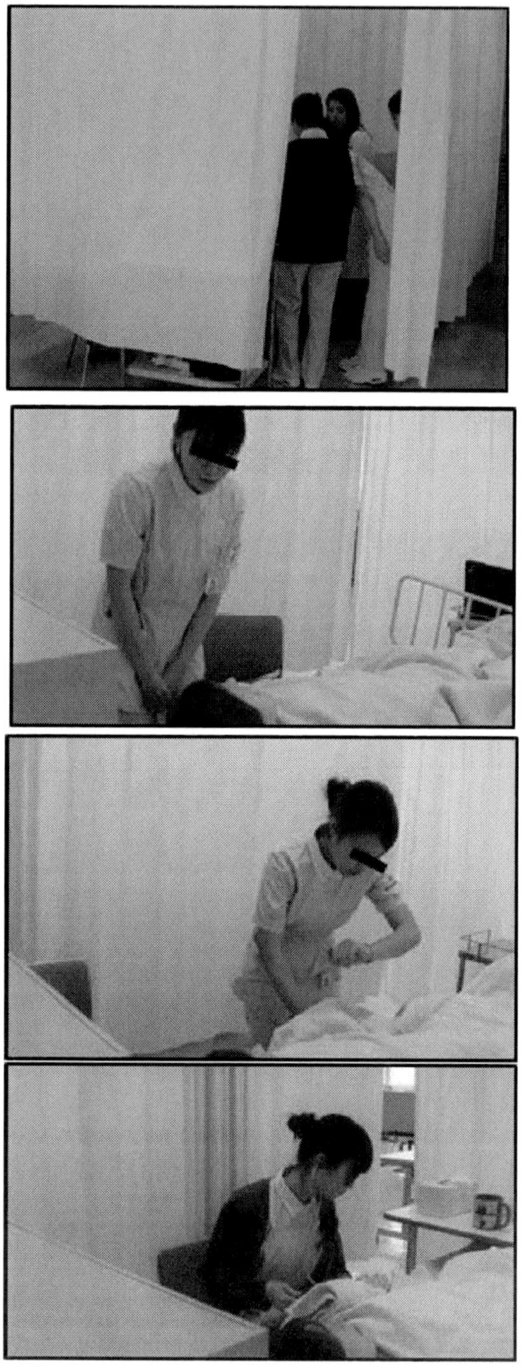

Figure 5. Simulation training flow. (a)Explanation of the simulated hospital room and instructor. (b,c) Participant observes the simluated patient to assess their condition. (10min/person) (d) Report patient condition using SBAR to the instructor (3-5min/person)

C. Survey tool

The survey tool assessed participants' opinions of the e-learning system (including amount of learning material, difficulty of questions, visibility, and ease of operation) and its effect on their desire to learn using a five-point Likert scale (extremely dissatisfied, slightly unsatisfied, neither satisfied nor dissatisfied, slightly satisfied, and extremely satisfied) and a free-comment section.

Desire to learn and factors influencing the successful implementation of learning were assessed using the four-subscale Instructional Materials Motivation Survey (IMMS), which was created based on the ARCS model [14]. This model contains four components: attention, relevance, confidence, and satisfaction. The IMMS, comprising 36 items, was designed to measure individuals' reactions to self-driven teaching materials, such as e-learning. A five-point Likert scale was used to rate each item, with higher scores indicating a more positive opinion. The IMMS includes items expressed negatively, which are reverse scored. In other words, before adding these answers to the total score, items given a rating of 5 were converted to have a rating of 1, with 4 being converted to 2, 3 remaining the same, 2 to 4, and 1 to 5.

After completing the entire training sequence, participants who gave consent were interviewed to express their thoughts regarding e-learning and simulation, and to determine the influence of simulation training on learning effectiveness and the desire to learn. All participants were given feedback about their performance after the simulation.

D. Study Design

Participants were divided into two groups by random assignment. The first group was simulation-only. After undergoing a brief orientation, participants completed a simulated patient visit according to the plan above. The clinical interaction was recorded on videotape and the video reviewed and scored according to a checklist.

Participants in the blended learning group, started with a pre-test to evaluate their approach to clinical evaluation of a patient. The pre-test was conducted as an integral part of the e-learning program using Moodle. Following the e-learning component, blended learning group then did the simulation, in the same manner as the simulation group. Following simulation training, participants in blended learning group completed a post-test using Moodle.

E. Statistical analysis

Test scores, times, and times of simulation are expressed as means ± standard deviations. The Wilcoxon signed rank test was used to assess differences in test scores before simulation and after. Differences in test scores were evaluated using the Mann-Whitney U test. For the questionnaire, scores are expressed as means and median values (quartile range). The free-response descriptions are listed as written by participants. All statistical analyses were performed using IBM SPSS Statistics 22 (IBM Corp., Armonk, NY). Analysis p-value<.05 is considered significant.

F. Ethical Considerations

This study was conducted with the approval of the medical ethics committee of Nagoya University, Japan. The survey was approved in June 2013 (No. 2013-0049). Subjects' freedom to withdraw from the study and protection of anonymity were explained both verbally and in writing. Written informed consent for this study was obtained from each participants.

III. RESULTS

A. Research Participation

A total of 46 junior nursing students participated in the study. Of these, 34 completed blended learning, e-learning and simulation, and 25 completed the questionnaire. The e-learning post-test was completed by 17. Interviews were completed for four participants. Simulation training only was completed by 12 participants who lacked the time to complete the e-learning.

B. E-learning

Correct answers on the e-learning pre-test were given by was 91% ± 8.5%. The correct answer rate for the measuring site of temperature and pulse, and the side effects of chemotherapy were 74%, 69%, and 74%, respectively. Participants took a mean of 22 ± 9 min to complete the e-learning.

Among the post-test participation group (n=17), the correct answer rate on the e-learning post-test was 99% ± 1.2%, which is significantly higher than the correct answer rate on the e-learning pre-test (87%±5.2%, p<.05), and reflects the effectiveness of e-learning. Participants who wrote 'free-form' comments often said that the e-learning was more likely to contain lists of words or wrong answers. In the e-learning post-test, the comments typically reflected the main points to be learned and described specific content, such as physical findings and diagnostic tests. Assessment of the patients' condition was also mentioned, in addition to recommendations about implementation of interventions based on patient care needs and priorities, and relevant patient education and teaching. The time needed for the e-learning post-test was 14 ± 4 min, significantly shorter than the time needed for the pre-test. The patients' vital signs in the e-learning post-test were different from those in the pre-test (Table II). The correct answer rates according to content are shown in Table III. Participants answered almost questions correctly and the answer rates are higher than on the pre-test.

TABLE II. PRE- AND POST-TEST ANSWER RATE AND TIME (N=17)

Contents	Pre-test	Post-test
Correct answers (mean ± standard deviation)	87% ± 5.2%	99% ± 1.2%
Time (min) (mean ± standard deviation)	24 ± 6	14± 4

TABLE III. RESULTS OF E-LEARNING PRE-TEST (N=17)

Number	Contents	Pre-test	Post test
1	The rules for entering a hospital room	100%	100%
2	Explanation of implementation	92%	100%
3	Correct site to measure temperature	74%	100%
4	Nomal range of body temperature	97%	100%
5	Correct site to measure pulse	69%	94%
6	Nomal range of pluse rate	95%	100%
7	Correct position for measurement of blood pressure for post- breast cancer operation patient	74%	97%
8	Methods for blood pressure measurement	92%	100%
9	Nomal range of blood pressure	92%	100%
10	Nomal range of respiratory rate	97%	97%
11	Side effects of chemotherapy	74%	74%
12	Communication with a vomiting patient	92%	100%
13	Check points for patient receiving chemotherapy	74%	100%
14	Communication with a constipated patient	92%	100%
15	Physical examination of the abdomen	92%	100%
16	Explaination of findngs releted to the patient's condition	91%	97%

C. Simulation

According to the checklists, all participants in the blended learning group were able to greet the patient upon entering the room, measure the patient's pulse, blood pressure, respiratory rate, and temperature, interpret the measured values, and exit the room after a polite goodbye. Confirmation of oral intake, sleeping condition, excretion condition, and side effects of chemotherapy were performed by 28, 21, 25, and 30 participants (n=34), respectively. Only two participants used information obtained from actually touching the patient, such as auscultation and palpation. Comparatively, participants in the simulation group were able to greet the patient upon entering room and exit the room after a polite goodbye. However, performance of other skills such as the side effects of chemotherapy and excretion condition were lower than for participants in the blended learning group. No participant in this group performed auscultation or palpation (Table IV). After feedback about simulation from the instructor, all participants understand the importance of physical examination such as auscultation and palpation, and included these modalities after the feedback.

The results of SBAR communication were different between the two groups. Participants in the blended learning group were more likely to make recommendations regarding "lack of balanced nutrition" and "risk for infection", as "consultation can help ensure that a diet plan is enacted", "proposal to have an antiemetic administered", " Initiate hand washing", "Wearing mask", and " Initiate mouth washing ", while participants in the simulation group were more likely to suggest issues such as "Anxiety" and "follow-up examination".

TABLE IV. SIMULATION CHECK LIST RESULTS

Contents	Number (%)	
	Blended learning group (n=34)	Simulation group (n=12)
-Greet the patient upon entering the room	n=34(100%)	n=12(100%)
-Measure the patient's pulse, blood pressure, respiratory rate, and temperature, and then interpret the measured values	n=34(100%)	n=7(58%)
-Exit the room after a polite goodbye	n=34(100%)	n=12(100%)
-Side effects of chemotherapy	n=30 (88%)	n=2 (17%)
-Confirmation of oral intake	n=28 (82%)	n=3 (25%)
-Excretion condition	n=25 (74%)	n=2 (17%)
-Sleeping condition	n=21 (62%)	n=3 (25%)
-Auscultation and palpation.	n=2 (5.9%)	n=0 (0.0%)

D. Questionnaire and Interview Results

Evaluation of e-learning and simulation: The results of the questionnaire assessing the e-learning characteristics are shown in Table V. Participants scored a value of 3 for all questionnaire items. In the free-response answers, one participant reported that the number of questions and difficulty of the e-learning was "exactly the right number and difficulty." Another reported that it was "a little too much to do in my spare time." Regarding visibility and operability, which had the most free-response answers, participants listed various concrete problems, such as "difficult to see," "difficult to use on a cellphone," "froze mid-answer," and "difficult to enter sentences on a smart phone."

In the interviews, participants reported the following on the simulation: "It was interesting, so I would like to use this method in classes as well"; "Because it was done in the e-learning, I could imagine it"; and "because I knew what was going to be implemented beforehand [owing to the e-learning], I was able to do it." One participant reported, "Although I understood the content, I became nervous and forgot it [during the simulation]." Additionally, participants reported contrasting views, such as "I was able to reconfirm my knowledge with the test" and "It was unnecessary because I was able to understand from the feedback during the simulation.

Influence on the desire to learn: Table VI shows the results for responses to all IMMS items. Item 15, "The pages of this lesson look dry and unappealing (reverse)", 33 "The content of this lesson will be useful to me" and 34 "I could not really understand quite a bit of the material in this lesson" (included in the Attention, Relevance and Confidence components), were scored at more than 4.5 points. In contrast, items 1 "When I first looked at this lesson, I had the impression that it would be easy for me" and 9 "There were stories, pictures, or examples that showed me how this material could be important to some people" (included in the Confidence and Relevance components), had a low median score (3 points). In the free-response answers, a number of participants expressed wanting more developmental content included in the e-learning. The participants' comments suggested a desire to use the e-learning system in the future, such as "In practice, I felt like I would be able to use it"; "I want to do normal exercises as well as e-learning and simulation"; and "As long as you know how it works, I think you could practice simulation similar to this with a classmate."

TABLE V. QUESTIONNAIRE ASSESSING E-LEARNING CHARACTERISTICS (N=34)

Contents	Mean, Median; Quartile range
Number of questions	2.9, 3; 3-3
Time taken	2.9, 3; 3-3
Difficulty	2.8, 3; 3-3
Readability	3.3, 3; 3-4
Character size	3.4, 3; 3-4
Screen design	3.7, 3; 3-3
Operability	3.0, 3; 3-3

IV. DISCUSSION

A. *Effects of Blended Learning*

Although simulation is an effective learning method and typically provides a greater sense of understanding, it can cause significant mental strain, which requires learning support. We developed a pre-simulation e-learning system to provide such support. The results showed that all participants performed the greeting, even if it was their first experience with simulation. Compared to the results of the simulation only group, using the e-learning system helped participants to assess vital signs, and complete physical assessment based on clinical data. This success can be attributed to participants' completion of the skill building and mock training exercises via e-learning system before the simulation. The e-learning program enabled participants to imagine the content of the simulation and thus integrate relevant knowledge and skills into their simulation experience.

However, participants did not completely master auscultation and palpation, according to information obtained from the interviews [15]. Therefore, the e-learning items related to this content may need to be modified. Specifically, the e-learning system can be modified such that, when repeatedly used, items that have been completely learned are excluded, enabling users to focus on items that were not completely mastered. Such a modification allows for the delivery of deliberate and effective training aligned with the learning objectives of nurse education, referred to as deliberate practice [16] [17]. In this study, students' ability to explain the side effects of chemotherapy did not increase the correct answer rate. Thus, the contents should be revised to be more effective. Learning self-management has been identified as a key factor for continued education via e-learning [7]. To succeed in remote self-learning, participants may need to receive explanations of the necessity of post-training tests in acquiring knowledge related to medical interviews, auscultation, and palpation, which are areas that remained challenging after the simulation training.

The results of SBAR communication were different in the two groups, and suggest that participants using e-learning decided on a recommendation based on information gathered from reading the patient's expressions and other situational and specific factors, while participants in the simulation group made their decisions based on general information.

B. *Effects of E-learning and Training on Motivation to Learn*

The ARCS model is a method of organizing factors that influence the motivation to learn [18], described by Suzuki as follows: A) Attention is captured. 'This looks interesting—there's something to this", R) Next, one realizes the relationship to themselves; knowing the learning task and realizing that 'it looks rewarding and it is related to my values.' Not only is the future value of the task significant, but also enjoying the learning process is valued. However, even if one finds significance in their learning, one can lose motivation when one recognizes that there is low possibility to accomplish the learning goal. C) On the contrary, if one has successful experiences in the first learning stages and can associate the experience with the endeavor, leading to the perception that 'I can manage it,' confidence is facilitated. S) Satisfaction: If one can feel fulfillment after looking back on the learning process and its accomplishment, it then leads to motivation to learn." In the present study, participants recognized e-learning as "having an important relationship to themselves," and they utilized this recognition as motivation to learn. To improve learners' confidence, we propose the following strategies: 1) Share evaluation criteria and allow learners to tackle tasks with a prediction of their possibility for success in mind; 2) adjust the difficulty of the system so that they can have a meaningful and successful experience;

TABLE VI. RESULTS OF THE INSTRUCTIONAL MATERIALS MOTIVATION SURVEY (N=34)

Items	Contents	ARCS	reverse	Mean, Median; Quartile
2	There was something interesting at the beginning of this lesson that got my attention.	A		3.5, 4; 4-4
8	These materials are eye-catching.	A		3.7, 3; 3-5
11	The quality of the writing helped to hold my attention.	A		3.5, 3; 3-4
12	This lesson is so abstract that it was hard to keep my attention on it.	A	○	4.2, 4; 4-5
15	The pages of this lesson look dry and unappealing.	A	○	4.5, 5; 5-5
17	The way the information is arranged on the pages helped keep my attention.	A		3.7, 3; 3-4
20	This lesson has things that stimulated my curiosity.	A		4.1, 4; 4-5
22	The amount of repetition in this lesson caused me to get bored sometimes.	A	○	4.1, 4; 4-5
24	I learned some things that were surprising or unexpected.	A		3.6, 4; 4-4
28	The variety of reading passages, exercises, illustrations, etc., helped keep my attention on the lesson.	A		3.6, 4; 4-4
29	The style of writing is boring.	A	○	4.1, 4; 4-5
31	There are so many words on each page that it is irritating.	A	○	4.4, 5; 5-5
6	It is clear to me how the content of this material is related to things I already know.	R		3.9, 4; 4-4
9	There were stories, pictures, or examples that showed me how this material could be important to some people.	R		2.9, 3; 3-3
10	Completing this lesson successfully was important to me.	R		4.3, 4; 4-5
16	The content of this material is relevant to my interests.	R		4.0, 4; 4-4
18	There are explanations or examples of how people use the knowledge in this lesson.	R		4.1, 4; 4-5
23	The content and style of writing in this lesson convey the impression that its content is worth knowing.	R		3.6, 4; 4-4
26	This lesson was not relevant to my needs because I already knew most of it.	R	○	4.2, 4; 4-5
30	I could relate the content of this lesson to things I have seen, done, or thought about in my own life.	R		4.1, 4; 4-5
33	The content of this lesson will be useful to me.	R		4.6, 5; 5-5
1	When I first looked at this lesson, I had the impression that it would be easy for me.	C		2.8, 3; 3-3
3	This material was more difficult to understand than I would like for it to be.	C	○	4.1, 4; 4-5
4	After reading the introductory information, I felt confident that I knew what I was supposed to learn from this lesson.	C		3.4, 3; 3-4
7	Many of the pages had so much information that it was hard to pick out and remember the important points.	C	○	4.1, 4; 4-5
13	As I worked on this lesson, I was confident that I could learn the content.	C		3.3, 3; 3-4
19	The exercises in this lesson were too difficult.	C	○	3.3, 3; 3-4
25	After working on this lesson for awhile, I was confident that I would be able to pass a test on it.	C		3.3, 3; 3-4
34	I could not really understand quite a bit of the material in this lesson.	C	○	4.6, 5; 5-5
35	The good organization of the content helped me be confident that I would learn this material.	C		4.0, 4; 4-4
5	Completing the exercises in this lesson gave me a satisfying feeling of accomplish¬ment.	S		4.0, 4; 4-4
14	I enjoyed this lesson so much that I would like to know more about this topic.	S		3.8, 4; 4-4
21	I really enjoyed studying this lesson.	S		3.9, 4; 4-4
27	The wording of feedback after the exercises, or of other comments in this lesson, helped me feel rewarded for my effort.	S		3.9, 4; 4-4
32	It felt good to successfully complete this lesson.	S	○	4.0, 4; 4-4
36	It was a pleasure to work on such a well-designed lesson.	S		4.1, 4; 4-5

and 3) provide opportunities and feedback that regulate learning, thus encouraging them to be aware that they can achieve their goal by themselves [19]. A number of participants commented that the training could be implemented as peer training and skill review. Repetitive practice through sharing stories with peers and skill review have been shown to help learners acquire self-confidence [20]. Further consideration of task difficulty will also facilitate continuous learning in this context.

C. Determinants of Successful E-learning

E-learning relies on independent learning, which means that there is an inherent risk of dropouts. Learning objectives were designed to help foster learning motivation and created teaching materials relevant to real-world clinical settings. In this study, the item "The content of this lesson will be useful to me" was rrated by a mean of 4.6 points, indicating that blended learning sustained participants' motivation. Additionally, the item "I could not really understand quite a bit of the material in this lesson" was reverse scored, also a mean of 4.6 points. The difficulty of blended learning and acquisition of confidence through simulation may be factors that influenced participants' motivation. The structure of the blended learning needs to be improved such that it can help participants gain confidence in their ability to perform the learning exercises and promote self-learning through repetitive practice and detailed explanations for items identified as weaknesses.

1) Operability is another factor that may have influenced participation. We expected this tendency in designing the system, and had tested the teaching material on smartphones prior to the survey. However, the interview results showed that when smartphones were used, the participants reported longer working time and poorer operability. This finding coincides with previous reports that the physical environment for online learning, such as Internet and terminal devices, influences compliance with e-learning [21]. For example, poor video teaching materials can reduce the motivation to learn [22]. Future research is needed to investigate additional factors related to operability, the differential effects of computers and smartphones, and the effects of the tendency to use smartphones for mobile learning, which is common among nursing students.

2) An interactive digital simulator for problem solving and clinical reasoning, both of which require physical assessment skills, was developed in Europe [23]. The digital simulator was developed for medical students and clinicians and uses virtual patients. Problem solving and clinical reasoning are needed in clinical nursing practice [24]. Future studies are needed to develop learning materials for nurses to acquire these required competencies.

V. Conclusion

These results show that e-learning was partially effective to improving physical assessment by nursing students. Additionally, e-learning helped participants recognize the relevance of learning content to their professional practice. It is necessary to combine e-learning and simulation to improve physical assessment ability in a blended learning program. However, the results also suggest that e-learning is insufficient for augmenting skills of auscultation and palpation, suggesting that teaching materials related to these skills must be improved using simulation training. Further development of e-learning, to make it more functional and to ensure that it sustains motivation to learn will be necessary, and requires continued study and evaluation.

Acknowledgment

Part of this research was supported by Grant-in-Aid for Young Scientists (B), 25862108, from Japan Society for the Promotion of Science (JSPS), JAPAN.

References

[1] M.S. Yagi, N. Miura, and S. Tsuzuku, "Effects of a Blended Learning System on Improving the Physical Assessment Ability of Nursing Students," in Proc. eTELEMED 2017, Nice, France.

[2] J. K. Hayden, R. A. Smiley, M. Alexander, S. Kardong-Edgren, and P. R. Jeffries, "The NCSBN National Simulation Study: A longitudinal, randomized, controlled study replacing clinical hours with simulation in prelicensure nursing education," J. Nurs. Regul., vol. 5, no. 2, pp. S1–S64, 2014.

[3] P. Moreno-Ger, J. Torrente, J. Bustamante, C. Fernández-Galaz, B. Fernández-Manjón, and M. D. Comas-Rengifo, "Application of a low-cost web-based simulation to improve students' practical skills in medical education, " International Journal of Medical Informatics, vol. 79, no. 6, pp. 459-467, 2010.

[4] L. Bickley and P. G. Szilagyi, Bates' guide to physical examination and history-taking. Philadelphia, Pennsylvania: Lippincott Williams & Wilkins, 2012.

[5] M. Fordis, J. E. King, C. M. Ballantyne, P. H. Jones, K. H. Schneider, S. J. Spann, et al., "Comparison of the instructional efficacy of Internet-based CME with live interactive CME workshops: a randomized controlled trial," JAMA, vol. 294, no. 9, pp. 1043-1051, 2005.

[6] J. G. Ruiz, M. J. Mintzer, and R. M. Leipzig, "The impact of e-learning in medical education," Acad. Med., vol. 81, no. 3, pp. 207–212, 2006.

[7] C. Jones, R. Ramanau, S. Cross, and G. Healing, "Net generation or Digital Natives: Is there a distinct new generation entering university?," Comput. Educ., vol. 54, no. 3, pp. 722–732, 2010.

[8] J. Keller and K. Suzuki, "Learner motivation and e-learning design: A multinationally validated process," J. Educ. Media., vol. 29, no. 3, pp. 229–239, 2004.

[9] J. Daniel, "Making sense of MOOCs: Musings in a maze of myth, paradox and possibility," J. Interac. Media Educ., vol. 2012, no. 3, 2012.

[10] M. Dougiamas and P. Taylor, "Moodle: Using learning communities to create an open source course management system," World Conference on Educational Media and Technology, vol. 1, pp. 171–178, 2003.

[11] M. S. Yagi, K. Ikegami, Y. Shibata, S. Shiba, and M. Miyoshi, "The design of the story-based learning materials to imagine realistic clinical practice using goal-based scenario theory," Japan Journal of Health Professional Development, vol. 3, no. 2, pp. 69–76, 2015.

[12] C. Arthur, T. Levett-Jones, A. Kable, "Quality indicators for the design and implementation of simulation experiences: a Delphi study," Nurse Education Today vol. 33, no.11, pp.1357-1361, 2013.

[13] M. Leonard, S. Graham, D. Bonacum, "The human factor: the critical importance of effective teamwork and communication in providing safe care," Quality and Safety in Health Care, vol. 13, suppl 1, pp. i85-i90, 2004.

[14] J. M. Keller, Motivational design for learning and performance: The ARCS model approach. New York City: Springer Science & Business Media, 2009.

[15] J. H. Barsuk, W. C.McGaghie, E. R.Cohen, K. J.O'Leary, and D. B. Wayne, "Simulation-based mastery learning reduces complications during central venous catheter insertion in a medical intensive care unit," Crit. Care Med., vol. 37, no. 10, pp. 2697–2701, 2009.

[16] K. A. Ericsson, K. Nandagopal, and R. W. Roring, "Toward a science of exceptional achievement," Annals of the New York Academy of Sciences, vol. 1172, no. 1, pp. 199–217, 2009.

[17] M. H. Oermann, S. Kardong-Edgren, T. Odom-Maryon, B. F Hallmark, D. Hurd, N. Rogers, et al., "Deliberate practice of motor skills in nursing education: CPR as exemplar," Nurs. Educ. Perspec., vol. 32, no. 5, pp. 311–315, 2011.

[18] J. M. Keller, Motivational design for learning and performance: The ARCS model approach. New York City: Springer Science & Business Media, 2009.

[19] K. Suzuki, "On the Framework of Designing and Developing 'appealing instruction': The ARCS Motivation Model," The Japan Association for Educational Media Study, vol. 1, no. 1, pp. 50–61, 1995.

[20] K. M. Lundberg, "Promoting self-confidence in clinical nursing students," Nurse Educ., vol. 33, no. 2, pp. 86–89, 2008.

[21] C. Power, H. Petrie, V. Sakharov, and D. Swallow, "Virtual learning environments: another barrier to blended and e-learning," in International Conference on Computers for Handicapped Persons, Berlin and Heidelberg: Springer, pp. 519–526, 2010.

[22] M. S. Yagi, N. Kumagai, C. Funakoshi, and M. Masuda, "Development of e-learning material for sophomore nursing students to simulate clinical practice," The Association for Medical Education in Europe, 270, #4JJ16 (126905), 2016.

[23] Body Interact. http://bodyinteract.com/ [retrieved: February, 2017]

[24] P. E. Benner, P. L. Hooper-Kyriakidis, and D. Stannard, Clinical wisdom and interventions in acute and critical care: A thinking-in-action approach. New York City: Springer Publishing Company, pp. 10–11, 2011G. Eason, B. Noble, and I. N. Sneddon, "On certain integrals of Lipschitz-Hankel type involving products of Bessel functions," Phil. Trans. Roy. Soc. London, vol. A247, pp. 529–551, April 1955.

Comparative Analysis of Walking Gait Cycle between Healthy People and Walking Disabilities to Prevent Tripping Using Wearable Device and KINECT

Yoshitoshi Murata, Shohei Yoshida
Graduate School of Software and Information Science
Iwate Prefectural University Graduate School
Takizawa, Japan
e-mail: y-murata@iwate-pu.ac.jo, g231n034@s.iwate-pu.ac.jp

Takayuki Niinuma, Kazuhiro Yoshida
Faculty of Software and Information Science
Iwate Prefectural University
Takizawa, Japan
e-mail: g0311121@s.iwate-pu.ac.jp, kyoshida@ipu-office.iwate-pu.ac.jp

Abstract— **Elderly people, especially walking disabilities, have an increased risk of falling and consequently injuring themselves. They need to be prevented from falling to maintain their health because injuries from falling are a major reason for them to be hospitalized or placed in residential care. Motion capture systems are a key component to prevent falls. We comparably analyzed walking gait cycle between healthy people and walking disabilities using a wearable device (WD) and KINECT to detect warning signs of falls. In this paper, we experimentally clarify what signs are useful to prevent falls. We developed a gait monitoring device comprised of a smartphone application and a pair of shoes on which WDs are mounted to measure such warning signs, and proposed presentation formats for data measured by KINETC.**

Keywords-falls; trip; hemiplegia; gait; shoes; MS-KINECT.

I. INTRODUCTION

The authors comparably analyzed walking gait cycles between healthy people and walking disabilities using gait monitoring shoes on which wearable devices (WDs) were mounted [1].

As the percentage of elderly people in the populations is increasing around the world [2], the number of functionally impaired people, such as cerebrovascular patients who are paralyzed down one side, will also increase. These people have an increased risk of falling and consequently injuring themselves [3][4]. Falling down is one of the main reasons for them to be hospitalized or placed in residential care.

There are many studies on falls by elderly people. The World Health Organization Regional Office for Europe analyzed these studies and classified fall risks amongst elderly people by history of falls, age, gender, living alone, ethnicity, medicine, medical conditions, impaired mobility and gait, sedentary behavior, psychological status - fear of falling, nutritional deficiencies, visual impairments, and foot problems [4]. Stroke patients, such as those with cerebrovascular disease, especially are at a substantially high risk of falling [5][6][7][8]. Their higher frequency of falls is due to weak muscles, one-side paralysis, and downward-pointing toes. For people with impaired mobility and gait, tripping is a major cause of falls [9][10], so we focus on tripping in this paper.

Since weak muscles, one-side paralysis, and downward-pointing toes strongly appear in the movement of legs and feet, motion capture for them is a key component to analyze impaired mobility and gait, and useful to prevent tripping, and conducts therapy and rehabilitation of hemiplegia.

Here, we focus on extracting warning signs of tripping for walking disabilities, such as cerebrovascular patients. In this paper, a WD is mounted on a shoe to measure the acceleration and angle velocity, and Microsoft KINECT [11] is used to measure positions of each joint of the lower body.

We obtained output data of an acceleration sensor and gyroscope sensor in a WD, Sony Smart Watch 3, mounted on the front part of a shoe to estimate the kicking power and change of angle between a foot and the floor. We noticed that the angle velocity at the terminal stance and the angle at the terminal swing are clearly different for unimpaired subjects and walking disabilities such as stroke. Moreover, they clearly have different step lengths as measured by KINECT. We also developed a monitoring device comprised of a smartphone application and a pair of shoes on which WDs were mounted, and proposed using the side and top view formats to present data measured by KINECT.

After introducing related works in Section II, we consider how people trip on a flat floor in Section III. Different features between physically unimpaired students and walking disabilities such as stroke are extracted from measured data in Section IV. Gait monitoring shoes and monitoring application, and gait presentation format with a KINECT are introduced in Sections V and VI. Measuring and analyzing a walking gait for walking disabilities are described in Section VII. Finally, conclusions are summarized in Section VIII.

II. RELATED WORKS

In this section, we introduce motion capturing devices.

A. Sensor usage type

Weijun Tao et al. reviewed gait analysis technologies based on wearable sensors that were the accelerometer, gyroscope, electromagnetic tracking system, magneto-resistive sensors, flexible goniometer, sensing fabric, force sensor, and so on [12]. They mentioned that fall risk estimation is an important application of gait analysis using wearable sensors. However, they did not describe about motion of gait for elderly people or walking disabilities.

Stacy J. Morris Bamberg et al. developed a prototype shoe in which several kinds of wearable sensors, such as accelerometer, gyroscope, force sensor, bidirectional bend sensor and so on [13]. The calibrated sensor outputs were almost same as results obtained simultaneously from a biological motion measuring equipment. They calculated the maximum pitch (angle between the shoe sole and floor at the toe-off timing), minimum pitch (angle between the shoe sole and floor at the heel-strike timing), the stride length from output of accelerometers and gyroscopes. They also compared the maximum pitch, minimum pitch and stride length between the healthy gait and parkinsonian gait. There were differences on mean value of calculated data between the healthy gait and parkinsonian gait. However, considering standard deviation of calculated data, such differences were small. They also did not measure and analyze motions of gait for elderly people or walking disabilities.

Farzin Dadashi et al. measured motion of gait for many elderly people with shoe-worn inertial sensors and provided normative values for a clinician to measure reference gait parameters [14]. They analyzed motion of gait and clarified the difference in gait parameters, such as the clearance between a shoe sole and floor, gait speed, stride length between males and females by considering the effect of age factors. However, their data did not show differences clearly between the male and female, and the effect of age factor. And, they did not investigate data for walking disabilities or analyze reasons for tripping.

Mourad Benoussaad et al. introduced a method to robustly estimate foot clearance during walking using a single inertial measurement unit (IMU) placed on the subject's foot [15]. In their paper, the foot clearance was the height of ankle from a floor. However, the toe clearance is more critical for tripping. And, they did not measure the toe clearance for walking disabilities such as stroke and analyze reasons for tripping.

B. Camera usage type

Vicon is one of the most famous companies in the motion capture industry. They can measure complex motions of joints in a body [16]. Vicon's system needs plural specialized video cameras, and know-how is needed to measure motions of joints. Thus, this system is too expensive for a small rehabilitation center or an individual to purchase and operate.

KINECT is one of motion capture devices distributed by Microsoft [11]. Since its price is a few hundred dollars, it is possible for small rehabilitation facilities to introduce it. There are many researches that use KINECT. Obdrzalek et al. compared the Kinect pose estimation with more established techniques relying on motion capture data [17]. They said that system such as Kinect has significant potential as a low-cost alternative for real-time motion capture and body tracking in health applications. We also used KINECT for a remote rehabilitation system of which content was a standing-up training [18]. In this paper, three kinds of view method, that are the front-view, side-view and top-view, were introduced to present a strain of the upper body.

III. Consideration of Tripping Factor

When the swing foot progression is unexpectedly obstructed, a trip occurs that leads to a forward rotation of the body and eventually might cause a fall.

Mourad Benoussaad et al. measured the minimum toe clearance (MinTC) to avoid tripping [15]. MinTC is a critical value to clear obstacles on the ground or floor. However, elderly people, especially those who have had strokes, sometimes trip on flat ground or floors, not obstacles. In this section, we consider reasons a person trips on flat ground or floors. We divide the normal walking gait cycle into eight phases the same as Weijun Tao et al. as shown in Fig. 1 [12]: (1) initial contact (heel-strike timing), (2) loading response, (3) mid-stance, (4) terminal stance (toe-off timing), (5) pre-swing, (6) initial swing, (7) mid-swing, and (8) terminal swing.

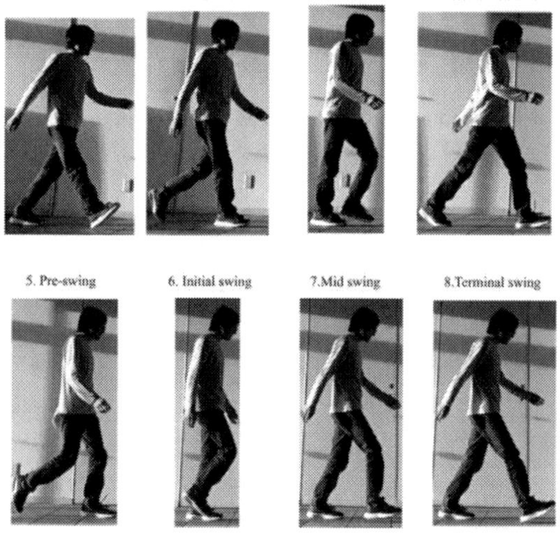

Figure 1. Normal walking gait cycle (See a right foot)

Most walking disabilities have weak muscles and are hard to raise their toe. They are at risk of three types of trips.

- Case 1: A toe touches the floor first instead of a heel at phase 1. Since phases 2-5 are skipped, the toe is dragged along the floor. When the dragging strength is stronger than the person's muscular power, he/she trips (Fig. 2(a)).
- Case 2: Kicking power of the front part of a foot is insufficient at phases 4 and 5 to raise the heel and toe up from the floor. In this case, a person does not swing but shuffles. When the frictional force between a shoe sole and the ground or floor is stronger than his/her muscular power, he/she trips (Fig. 2(b)).
- Case 3: A toe touches the floor due to it pointing down during the swing phases (5-8), and the knee goes further forward than the foot. When the dragging strength is stronger than the person's muscular power, he/she trips (Fig. 2(c)).

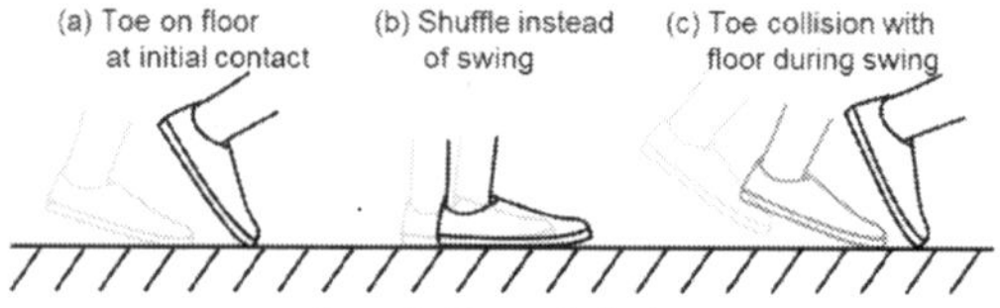

Figure 2. Cases of tripping

The above problems suggest that the kicking power at phases 4 and 5 and the angle between the foot and a floor are critical parameters.

IV. EXTRACTION OF WARNING SIGNS FOR TRIPPING

In this section, we experimentally investigate whether the kicking power at phases 4 and 5, and the angle between the foot and lower limb are critical parameters.

A. Experimental method

Since kicking power must be expressed as the angle velocity or the acceleration for the foot, we mounted a WD which had an accelerometer and gyroscope on the foot. In this experiment, we used Sony SmartWatch 3 as a WD which is mounted on the front part of a foot with Velcro tape as shown in Fig. 3. This mounting position was same as one in Farzin Dadashi's experiment [14]. The sampling rate was 40 msec.

We measure angle velocity for up and down directions of the front part of the foot (X axis of a 3D gyroscope). We also adopted a three-point moving average of the angle velocity to calculate the angle, because output values extremely change up and down. Therefore, the angle for X axis $Angle_{xn}$ at time t_n is calculated as follows.

$$Angle_{xn} = Angle_{xn-1} + \frac{t_n - t_{n-1}}{1000} \times \frac{G_{xn-1} + G_{xn} + G_{xn+1}}{3} \quad (1)$$

G_{xn} is the value of angle velocity for X axis at time t_n.

We investigated the measuring accuracy of Sony SmartWatch 3 using a slant rule as shown in Fig. 4. We measured data five times. Calculated angles vs. angles given by the slant rule are listed in Table I. These data showed calculated angles were so accurate. We noticed drift errors of a gyroscope that increase the value by 0.2 rad. /sec. during a WD sets on a flat floor. However, each measurement lasted less than 20 sec. Therefore, we think the effect of the drift error is negligible.

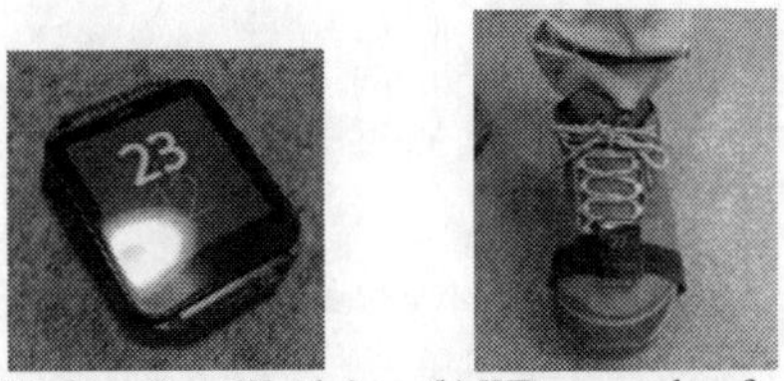

(a) WD: Sony SmartWatch 3 (b) WD mounted on foot
Figure 3. Measuring device and WD mounting method

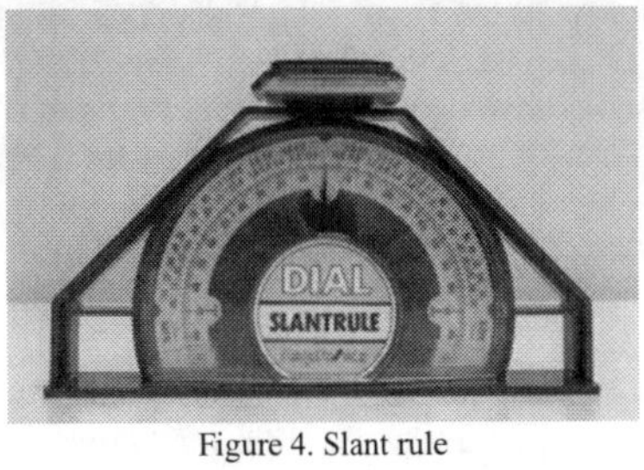

Figure 4. Slant rule

TABLE I. ACCURACY OF CALCULATED ANGLES

Given angle (degree)	Calculated angle (degree)	Standard deviation (degree)
+50	+49.00	0.45
+40	+39.17	0.74
+30	+28.92	0.51
+20	+19.58	0.63
+10	+9.21	0.58
0	0.21	0.15
-10	-9.72	0.58
-20	-19.47	0.34
-30	-30.70	0.56
-40	-40.57	0.41
-50	-50.54	0.53

We also measured the foot stride (FS) and recorded motions of subjects' knees, ankles, and feet by using KINECT. The UNIX time was introduced to synchronously measure data with a WD and KINECT.

We defined the FS as the maximum difference between positions of the right and left foot measured by KINECT as shown in Fig. 5. We experimentally looked for the height of KINECT to measure accurately. As the result, the height of KINECT is 75 cm. Moreover, we set the face angle of KINECT so that "+" markers on the display are superimposed on "-" markers on the floor to correct KINECT as shown in Fig. 6. From the bottom of the image in Fig. 6, these markers correspond to 2, 3, 4, and 5 m from KINECT. Since this picture is output of KINECT's video camera, left and right are reversed.

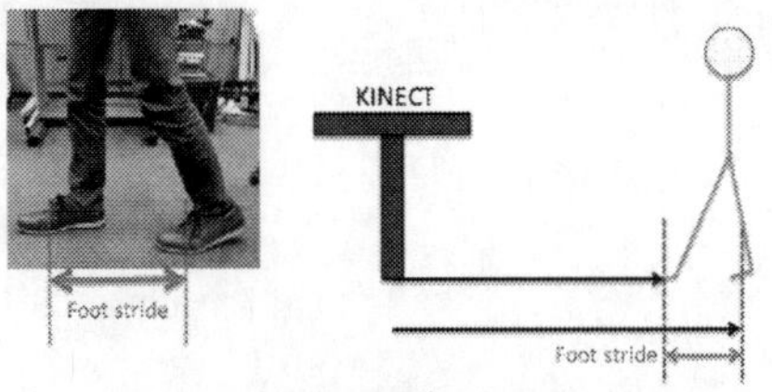

Figure 5. Definition of the foot stride (FS)

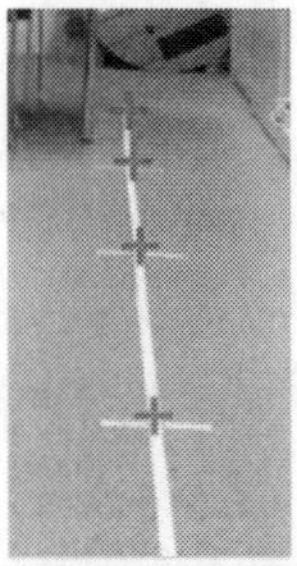

Figure 6. Pre-setting of KINECT

Fig. 7 (a) shows distance from KINECT for a left and right foot measured by KINECT. The blue line shows the distance between the left foot and KINECT, the red line shows the distance between the right foot and KINECT, and the green line shows the length for the left foot minus the length for the right foot. The minimum and maximum values of the green line very clearly correspond to FS for each step. Three-point moving average curves for measured data are shown in Fig. 7 (b) for reference. Circled parts correspond to movement of the feet from the mid-stance to terminal stance. Since the three-point moving average masks such feet motions, we decide to present a graph containing raw data.

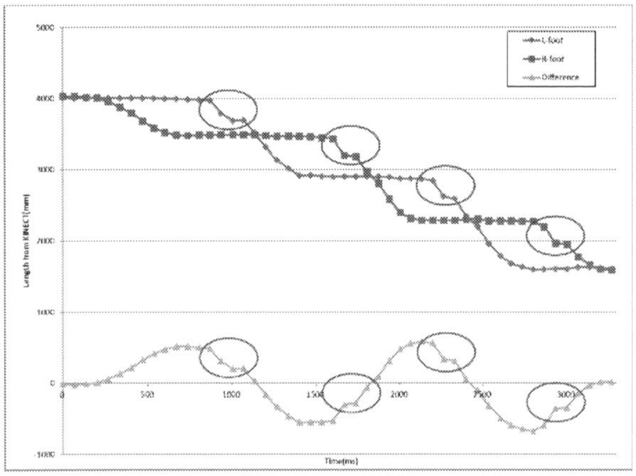

(a) Raw measured data

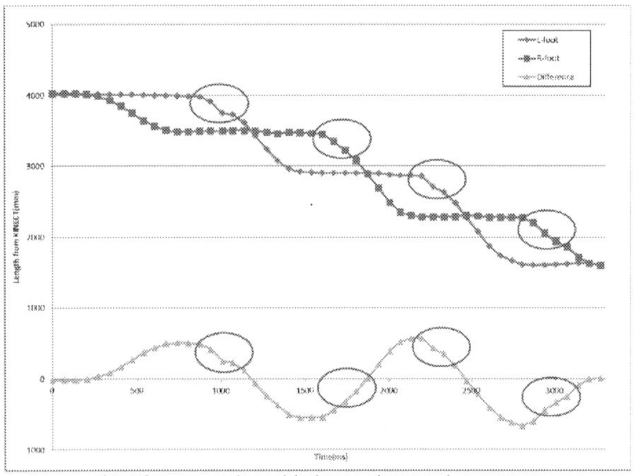

(b) After processing with three points moving average

Figure 7. Measured length between KINECT and the left or right foot

We experimentally evaluated the accuracy of measured FSs. Participants were eight unimpaired university students. They walked on three sets of stride markers of 30, 60, and 70 cm as shown in Fig. 8.

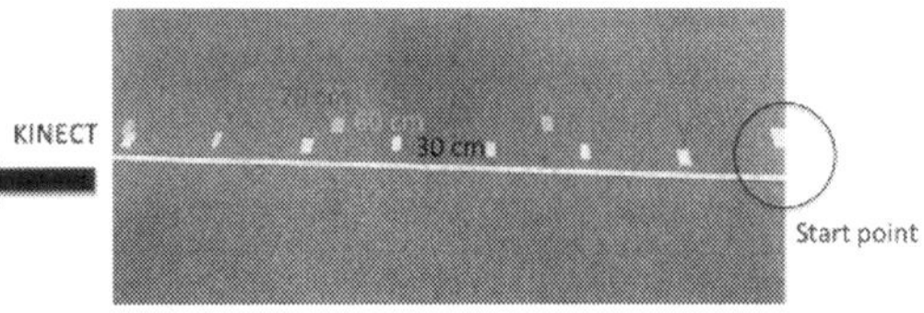

Figure 8. Three pitches markers for the accuracy of measured FS

Experimental results are shown in Table II. Since the range within which KINECT can definitely measure is between 1.5 and 4.5 m, average values in Table II are averages of absolute values for strides between them. Errors were less than 10%, and the standard deviations were less than 3 cm. We evaluate the errors within allowance.

TABLE II. ACCURACY OF MEASURED FS

Pitch (cm)	Average (cm)	Standard deviation (cm)
30	27.7	2.02
60	54.2	2.65
70	63.1	2.67

B. Measured data and consideration

We measured the acceleration, angle velocity and angle for five physically unimpaired students and three walking disabilities using a WD as same as the former sub-section. Every walking disabilities in this experiment had one-side paralysis, and trained periodically at a rehabilitation facility. Some of them used a wheel chair and could not walk by himself before training. They walked along a straight line to MS-KINECT. A WD was attached on the front part of foot on the paralysis side as shown in Fig. 9. We measured data for each patient two times.

(a) MS-KINECT set up in rehabilitation facility

(b) WD mounted on a foot of patient

Figure 9. Measurement environment in rehabilitation facility

Figs. 10 and 11 show examples of change of acceleration, angle velocity, and angle for a physically unimpaired student and a walking disability. Data for two steps are plotted.

Each flat period (roughly the center period) in these figures is when the entire shoe sole touched the floor; this period corresponds to phases 2 (loading response) and 3 (mid-stance). The reason that the value during this period is not zero is that the WD measures the angle between the front part of the foot and the floor, which depends on the person and shoe. Therefore, we reset this angle for the gait monitoring shoes described in Section V when the entire shoe sole touched the floor. This processing enables the WD to measure the angle between the back of the foot and the floor, and removes the drift error of the gyroscope. This value does not depend on person or shoe.

The maximum angle velocity at timing A means the kicking power from phase 4 (terminal stance) to 5 (pre-swing), and the minimum angle at timing B means the angle to the floor at phase 8 (terminal swing).

Lower angle velocity at A in Fig. 10 is about 420 deg./sec. On the other hand, higher angle velocity at A in Fig. 11 is about 250 deg./sec. Thus, a physically unimpaired student and a walking disability obviously differ in terms of gait. The walking disability clearly has weaker kicking power at phase 4 (terminal stance) than the physically unimpaired student.

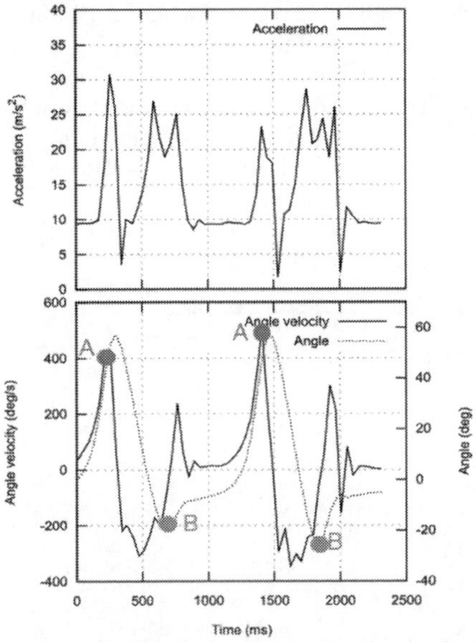

Figure 10. Changes of angle velocity, angle, and acceleration for physically unimpaired student

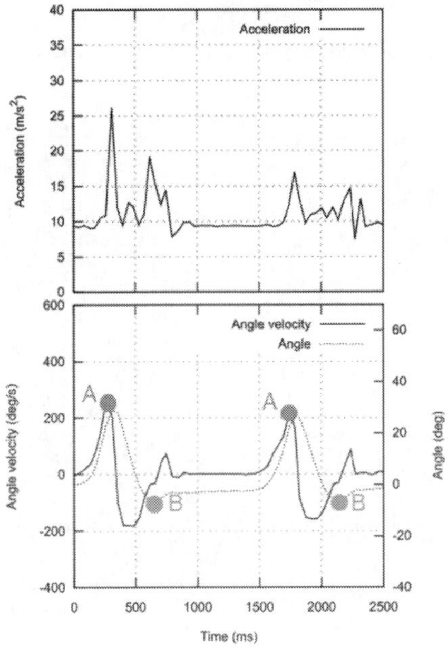

Figure 11. Changes of angle velocity, angle, and acceleration for walking disability

Higher angle at B in Fig. 10 is about -18degree. On the other hand, lower angle at B in Fig. 11 is about -8degree. Thus, a physically unimpaired student and a walking disability obviously differ in terms of the angle to a floor at phase 8 (terminal swing). This shows that it is difficult for a walking disability to raise his or her toe at the terminal swing phase.

The other hand, the acceleration basically changes corresponding to the angle velocity and angle. However, they have much noise, and their amplitudes are not stable.

Tables III and IV list the averages and standard deviations (SDs) of measured data for angle velocity at timing A and angle at timing B. The angle velocity at timing A is clearly different between unimpaired students and walking disabilities. There is a big difference between them in the angle at timing B, however, this value would have sometimes overlapped each other.

Table III. Angle velocity at the terminal stance

Participant	Average (deg./s)	SD (deg./s)
Student	509.36	18.91
Walking disability	342.06	86.52

Table IV. Angle at the terminal swing

Participant	Average (deg.)	SD (deg.)
Student	-17.76	8.02
Walking disability	-7.45	8.02

We also measured the FS using KINECT, and the cadence for a gait using a WD and KINECT. Table V lists the averages and SDs of measured data for the strides. In this paper, we define the cadence as the number of steps per minute. We estimated the cadence derived from an average of 10 intervals between one timing A and the next A, which were peak angle velocities of a step, when a WD was used. Estimated cadences are listed in Table VI.

There are clearly differences between unimpaired participants and walking disabilities in terms of the FS. FSs of walking disabilities are more than 10 cm shorter than those of unimpaired participants. On the other hand, the cadences of walking disabilities are slightly faster than those of unimpaired participants. Most physiotherapists said that FSs of elderly people, especially walking disabilities, are usually shorter than those of unimpaired people. These data prove what physiotherapists know experimentally.

TABLE V. FOOT STRIDE (FS)

Participant	Average (cm)	SD (cm)
Student	60.0	6.5
Walking disability	42.6	1.8

TABLE VI. ESTIMATED CADENCE

Participant	Average (steps/m)	SD (steps/m)
Student	46.4	5.0
Walking disability	49.0	6.2

On the basis of the results of these experiments, **we decided to adopt the angle velocity at the terminal stance to initial swing, angle between a foot and floor at the terminal swing, and average FS to detect warning signs of falls.** Section V introduces a pair of shoes and smartphone application to measure angle and angle velocity, and Section VI shows presentation formats for data measured by KINECT.

V. Gait Monitoring Device

A. Shoes

A WD has to be attached somewhere on a body during walking to detect signs of tripping to prevent a fall. A WD was attached to the front part of the foot in Section IV. However, it is difficult for a WD to firmly be set at this place for a long time because it is easily detached. Therefore, we studied which position is the best to detect the change of angle velocity for a foot and angle between a foot and floor. We attached WDs to a heel and a lower limb as shown in Fig. 12.

For this test, we used STEVAL-WESU1 by STMicro-electronics (see Fig. 13) as a WD instead of Sony SmartWatch 3. This wearable unit includes four sensors:

- 3D-accelerometer,
- 3D-gyroscope,
- 3D-magnetometer,
- MEMS pressure.

This device is 37 x 40 x 8 mm and weighs 9.6 g.

We inserted STEVAL-WESU1 into the heel of a shoe as shown in Fig. 12 (a) (details in the next sub-section). Angle velocity and acceleration data of STEVAL-WESU1 are sent to and processed by an Android smartphone. The sampling rate was 40ms. We adopt a three-point moving average to remove noise.

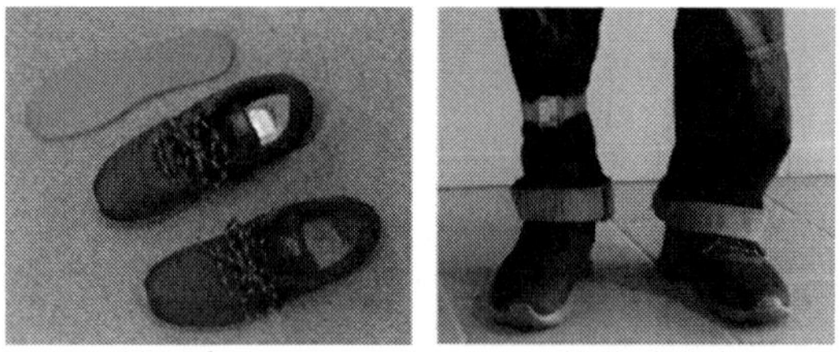

(a) Heel (b) Single limb

Figure 12. WD attaching position

Figure 13. STEVAL-WESU1 by STMicroelectronics

We requested three unimpaired students to walk with their normal gait. Since their data change was basically the same, graphs of one participant are shown in Figs. 14 and 15. Both plotted lines in Fig. 14 are similar in shape to those in Fig. 10. Timing A and B correspond to timing A and B in Fig. 10. Timing B in Fig. 14, which is the angle at the terminal swing, is shown more clearly than that in Fig. 10. On the other hand, timing C in Fig. 15 shows the kicking power from phase 4 (terminal stance) to 5 (pre-swing) is the same as timing A in Figs. 10 and 14. However, the angle at timing D in Fig. 15 is between not the foot and floor but a single limb and the vertical line to the floor. The plotted angle in Fig. 15 clearly shows a change of angle for the single limb.

As the result of this experiment, we decided that the heel was the best position to place a WD.

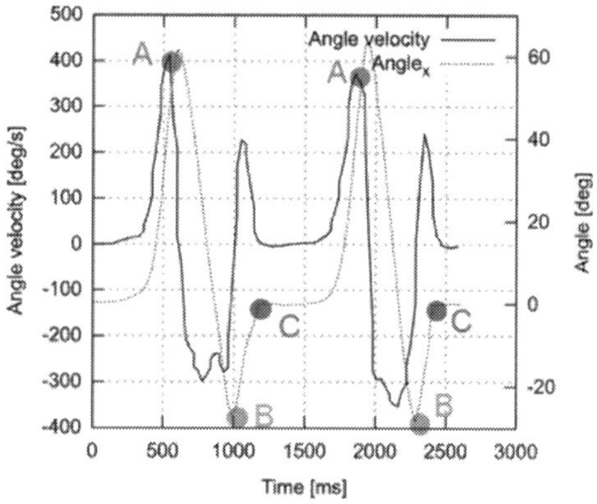

Figure 14. Angle velocity and angle data at heel in normal walk

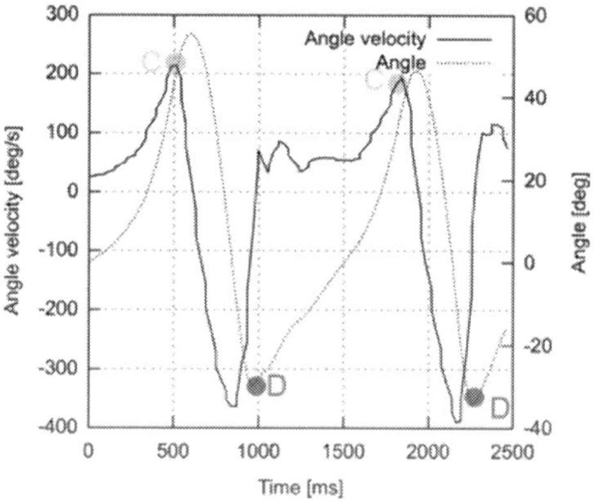

Figure 15. Angle velocity and angle data at single limb in normal walk

B. Monitoring application for smartphone

As described in former sub-section, we determined the heel of a shoe is the best place to measure angle velocity of the foot and angle between the back part of a foot and the floor. We inserted a WD (STEVAL-WESU1 by STMicroelectronics) into soles of both shoes. And, we also developed a gait monitoring application for Android smartphone which measures and stores the angle velocity and angle as shown in Fig. 12. The upper part shows ID of WD for the right and left shoe, and the lower part shows angle velocities at A in each step for right foot, angle at B in each step for right foot, angle velocities at A in each step for left foot, and angle at B in each step for left foot. In this application, direction of angle is turned. When these graphs were measured, a participant played a stroke patient who had a one-side paralysis for the right side of the body. Therefore, most strength of angle velocity at A for right foot were smaller than that for left foot. And, most amplitude of angle at B for right foot were smaller than that for left foot.

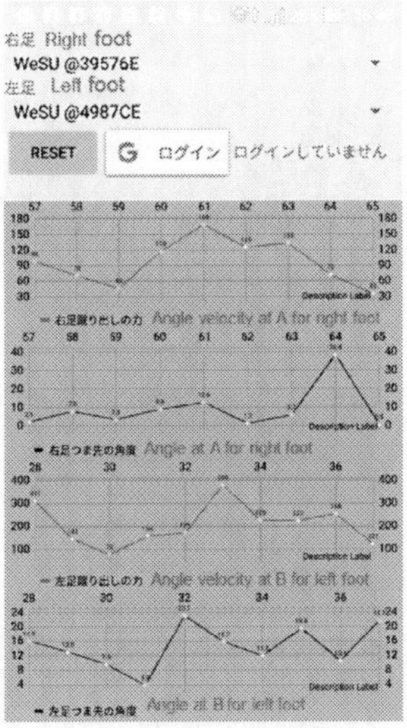

Figure 12. Gait monitoring application for Android

C. Measured data using gait monitoring shoes

We experimentally monitored the walking gait for two participants. They were unimpaired people. They walked and played the three types of trips in Fig. 2. Example measured data are shown in Fig. 13. Therefore, curves of angles in these graphs have different discontinuity to those in other graphs at the sole of a shoe touching a floor.

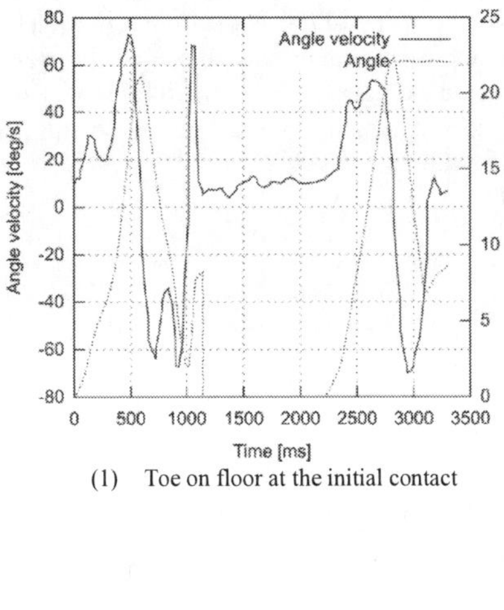

(1) Toe on floor at the initial contact

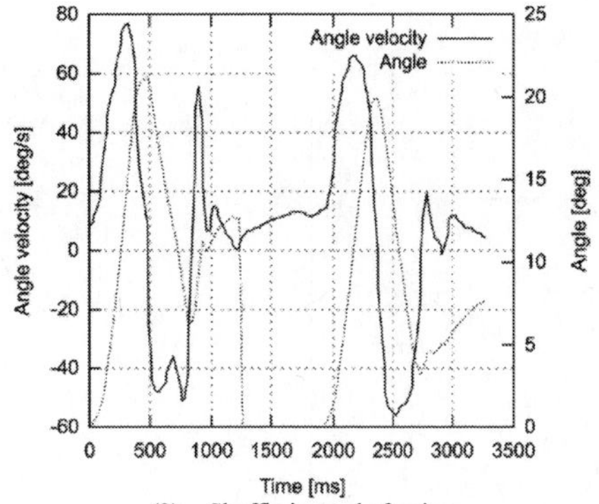

(2) Shuffle instead of swing

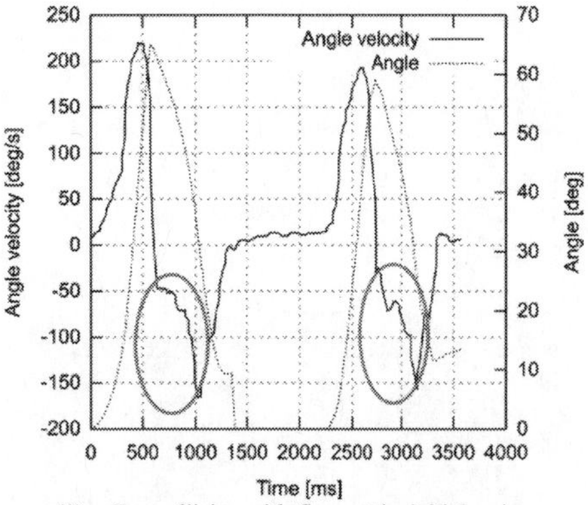

(3) Toe collision with floor at the initial swing

Figure 13. Example measured data for tripping with gait monitoring shoes

In (1); toe touching the floor first instead of a heel, and (2); shuffling, shapes of angle velocity resemble that of the normal walk shown in Fig. 10. However, maximum values of angle velocity and minimum angle in a cycle in Fig. 13 (1) and (2) are much smaller than those in Fig. 10. Their absolute minimum values are also much smaller than those in Fig. 10. This feature must show that when muscle strength is weaker, more trips occur. The red circle in Fig. 13 (3) shows this situation clearly. In the case of a normal walk, angle velocity rapidly decreases from the pre-swing to the initial swing. However, in (3), the angle velocity limply decreases on the way.

VI. GAIT PRESENTATION FORMAT FOR KINECT

The above gait monitoring device is useful to measure degrees of muscle power and the angle between the foot and floor. However, they have difficulty measuring the position of joints of the lower body such as the foot and ankle. Hence, we measured them using KINECT. Since we noticed that three kinds of view image (front, side, and top views) were useful to find out the strain condition of the upper body [18], we adopted these three view graphs for the gait cycle. Before measuring joints of the lower body for walking disabilities, we measured them for an unimpaired person to evaluate

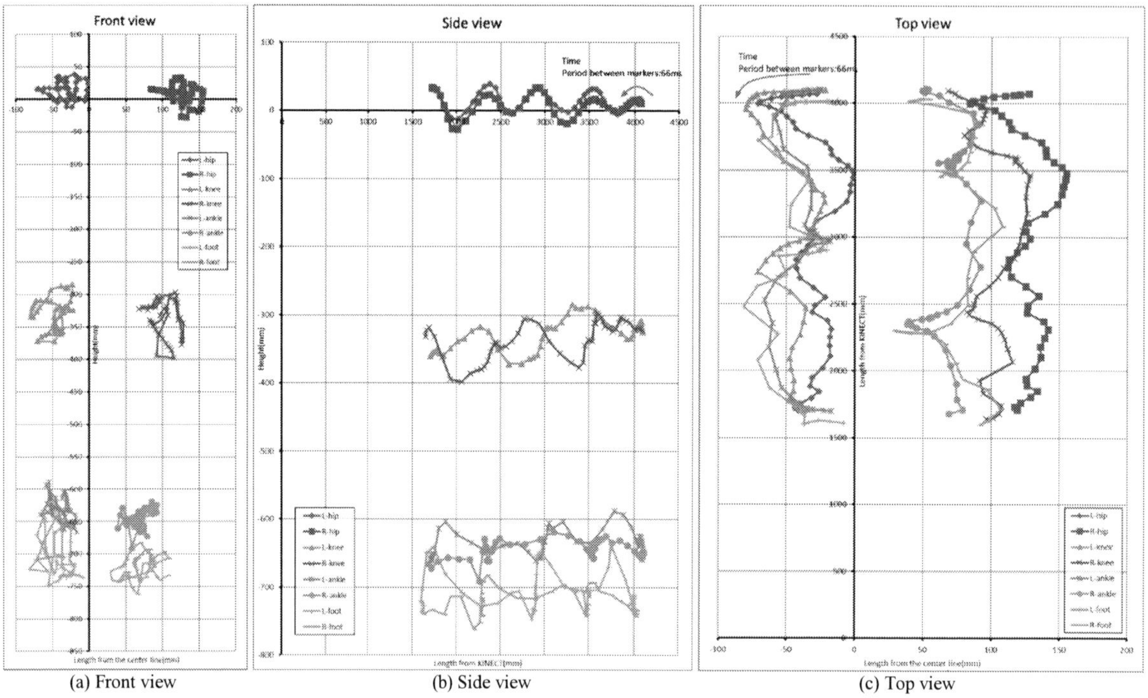

(a) Front view (b) Side view (c) Top view

Figure 14. Positions of joints on the lower body for natural gait

KINECT's measurement accuracy. The walking course was the line shown in Fig. 6. A participant walked in three ways:

(1) Natural gait: a participant puts one foot on either side of the center-line (Fig.14).
(2) On-line gait: a participant puts his/her foot on the center line (Fig.15).
(3) Circumduction gait: a participant moves his right foot naturally and exaggeratedly rotates his/her left-foot away from the center-line (Fig.16). Some hemiplegia patients move their palsied foot with this walking form.

In Figs. 14 to 16, the original position is the center of KINECT for each direction. Since KINECT is set 75 cm above the floor, height of the floor is -75 cm. Data for hips, ankles, knees, and the front part of feet are presented in these figures.

The front-view shows moving height ranges for each joint. However, it is impossible to detect the position of each joint of the basis of walking steps. On the other hand, the side and top views respectively show the change of each joint in the vertical and horizontal directions in accordance with walking. The change of the moving height range of hips is very clearly shown in all side view graphs. However, measured data for the heights of knees, ankles, and feet by KINECT did not change smoothly. Since lengths from KINECT for both feet were very accurately measured as shown in Fig. 7, we decided not to adopt the moving average processing. Hence, we will consider whether some processing should be adopted to smooth them.

A top view curve in each graph shows the change of length from the center line. From Figs. 14 and 15, measured data were shifted to the right when KINECT was used. However, left and right feet were put on the center line one after the other in the case of on-line walking (Fig. 15 (c)). This means that the measured length from the center line is basically accurate.

When a participant exaggeratedly rotated his/her left-foot away from the center-line in the circumduction gait, curves of his/her left knee, ankle, and foot in the top view clearly showed their motions as shown in Fig. 16 (c). Moreover, their curves in the side view showed that their motions in the vertical direction were bigger than those of the right foot. KINECT detected exaggerated motion with high accuracy.

One physiotherapist said that curves in the side and top views would be useful to instruct hemiplegia patients to walk more generally.

VII. Measuring and Analysing a Walking Gait for Walking Disabilities

We measured angle velocity at the terminal stance to initial swing, angle between a foot and floor at the terminal swing, and average FS using a proposed gait monitoring device and KINECT. Positions of joints on the lower body measured by KINECT were presented with the side and top

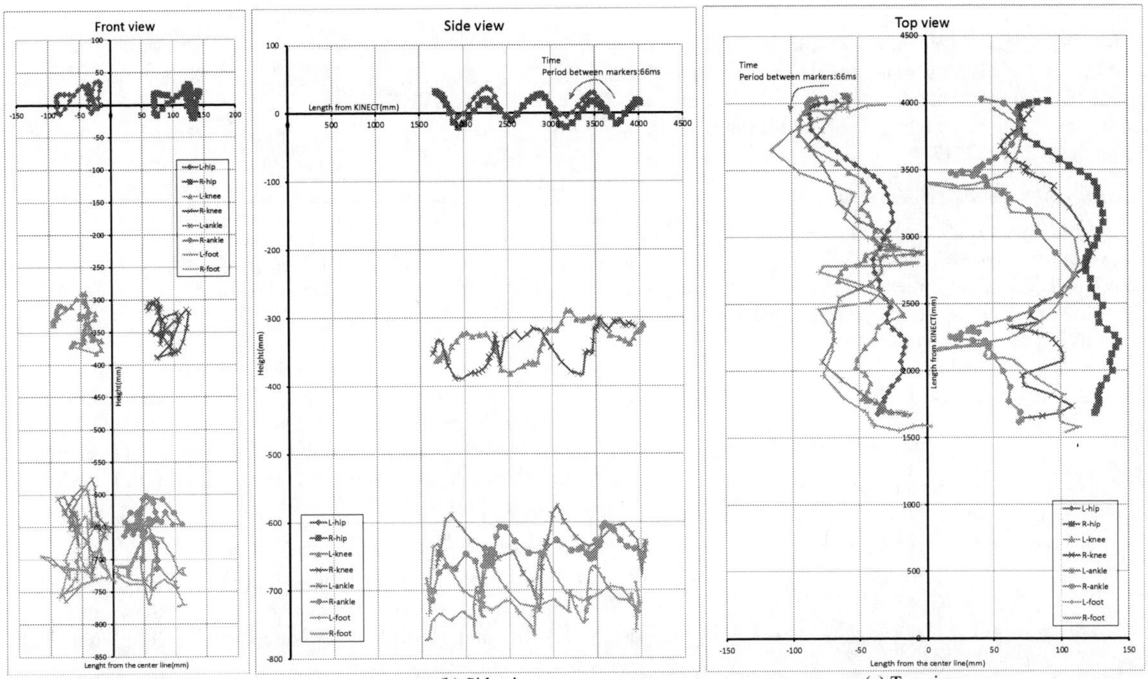

(a) Front view (b) Side view (c) Top view

Figure 15. Positions of joints on the lower body for on-line gait

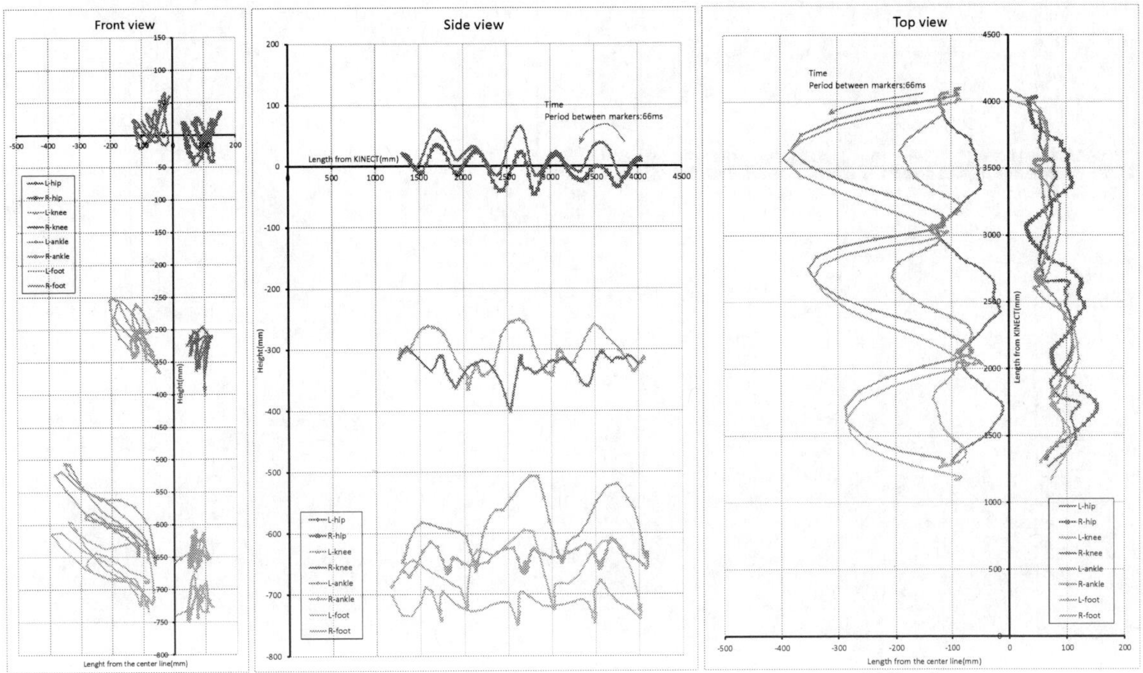

(a) Front view (b) Side view (c) Top view

Figure 16. Positions of joints on the lower body for circumduction gait

view formats. Since a gait monitoring device did not work well at the first measuring date, we measured four patients using KINECT only at that day (17/08/2017). We measured different four patients using a gait monitoring device at different date (23/08/2017).

A. *Measured by a walking gait device*

Measured data for four walking disabilities by a walking gait device are showed in Fig. 17. Profiles of patients are listed in Table VII. Since Patient A has paralysis in right-side, his right foot angle velocities at the pre-swing are lower than those of his left foot. His right foot angles at the initial contact are approximately equal to an average foot angle of walking disabilities listed in Table VI. However, those of his left foot

Table VII. Patient profiles measured by a gait monitoring device

Patient	Sex	Age	Symtom
A	Male	77	Stroke Right-side paralysis
B	Male	67	Stroke Left-side paralysis
C	Male	80	Cervical myelopathy Lumgago
D	Male	70	Quadriplegia Numbness in both shoulders and fingers

are changed so hard. Patient B has paralysis in his left-side. However, his physical strength is weak, and there is not clearly difference between his right and left foot. Patient C and D do not have any paralysis. Their physical strengths are a little weak, and their toe angles at the initial contact are basically low.

B. *Measured by KINECT*

Measured data for four walking disabilities by KINECT are showed with the side and top-view as shown in Fig. 18. Profiles and average FS for patients are listed in Table VIII.

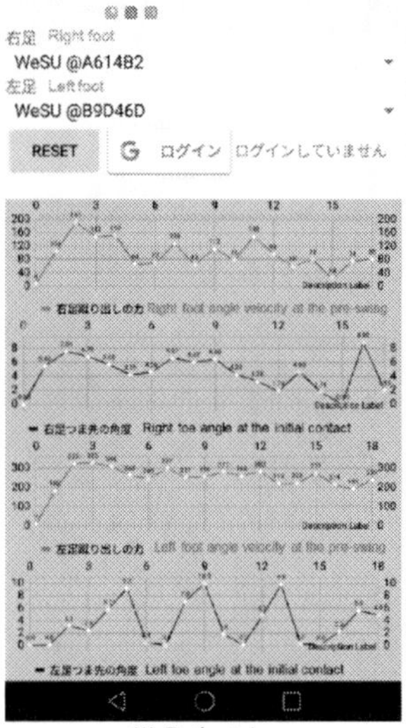

(1) Patient A

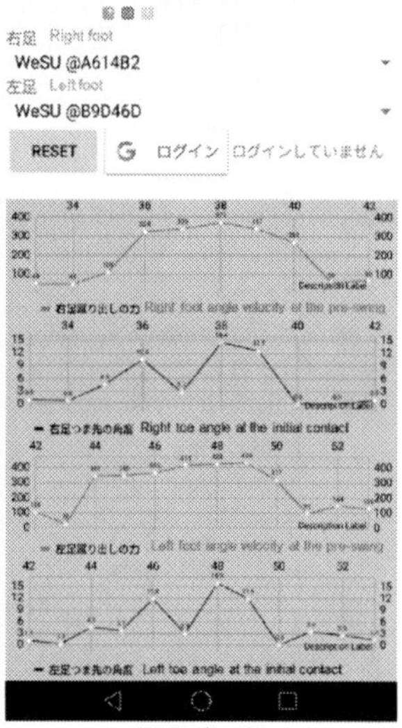

(2) Patient B

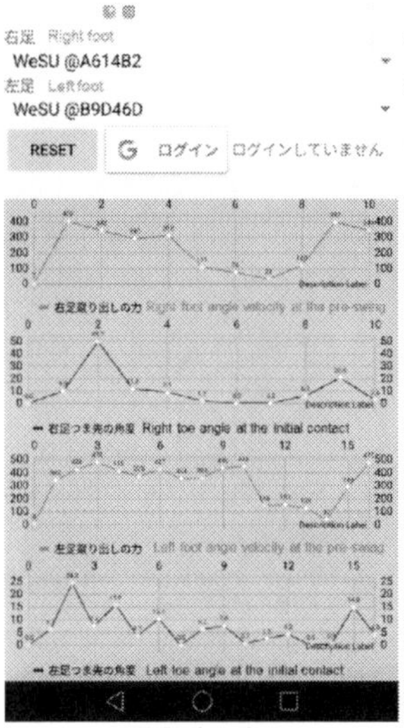

(3) Patient C

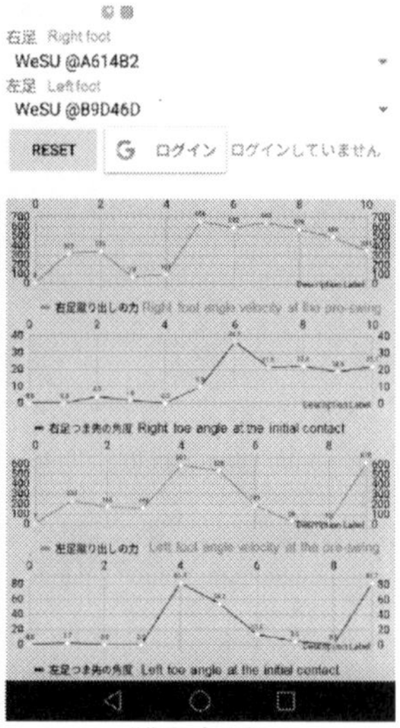

(4) Patient D

Figure 17. Measured data with a walking gait device

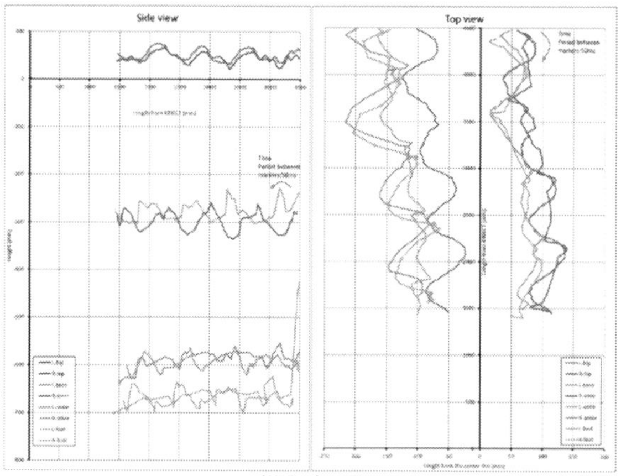

(1) Patient E

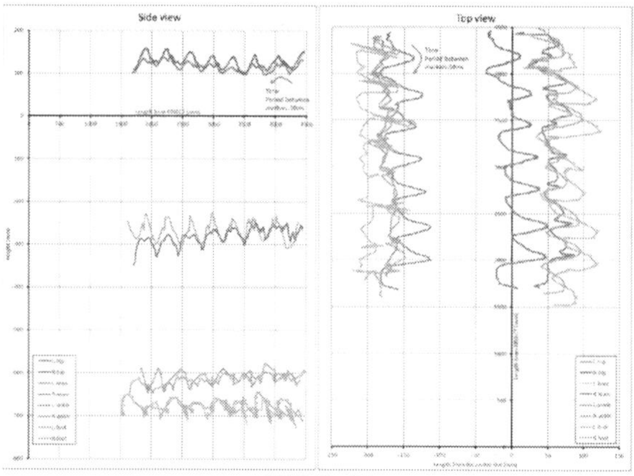

(2) Patient F

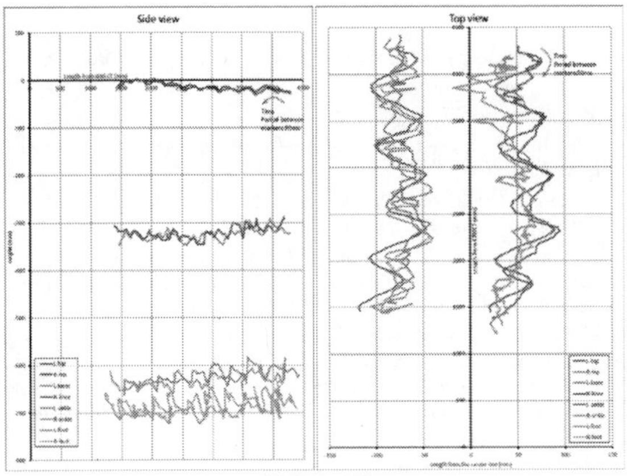

(3) Patient G

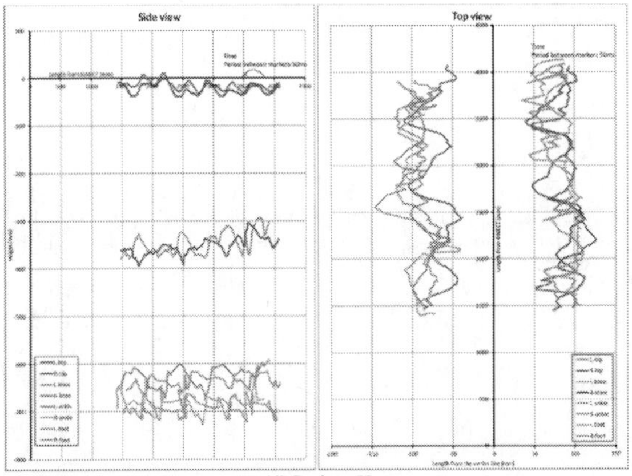

(4) Patient H

Figure 18. Side and top views for joints of walking disabilities

Table VIII. Patient profiles and foot stride (FS) measured by KINECT

Patient	Sex	Age	Symtom	Av. FS (mm)
E	Male	64	Stroke Left-side paralysis	370
F	Male	79	Stroke Right lower leg break	215
G	Female	64	Vertebral canal stenosis Both shoulder-ache	303
H	Male	73	Vertebral canal stenosis Left foot downward-pointing toe	367

The average FS for every patient are less than that of walking disability listed in Table V. These data correspond to what physiotherapists said they have a gait disturbance.

The top view curves in Fig. 18 (1) shows that Patient E walks in the circumduction gait for his left-side and has paralysis in his left-side. An average FS of Patient F is very shorter than others. The curve of FS is usually changed symmetry to cross points of both feet. However, his curve is not asymmetry as shown in Fig. 19. The reason is that he uses a cane with his left hand, he sends a cane first, and then his right foot near by the position of cane, and finally his left foot a little beyond his right foot in a gait cycle. Since he leans on a cane with his left hand, his right- hip is shifted to the left.

Side-view curves of Patient G's hip and knee in Fig.18 (3) show that her both hips and knees do not almost move up-and-downward. The chief physiotherapist in this rehabilitation facility says that she must have mastered this walking gait to avoid ache in her both shoulders. There are not any features for Patient H in Fig. 18(4).

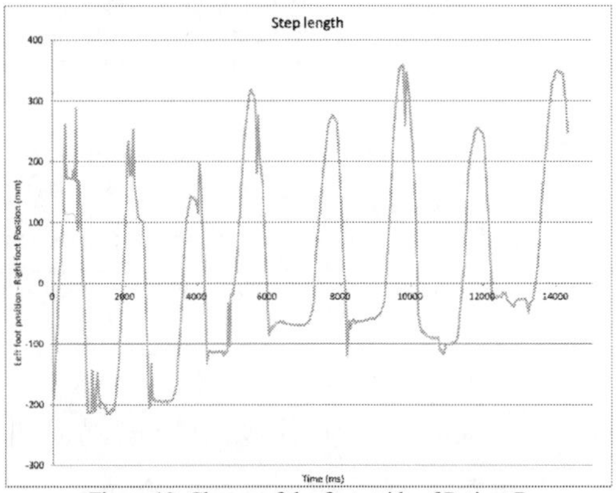

Figure 19. Change of the foot stride of Patient B

VIII. Conclusion

We comparably analyzed walking gait cycle between healthy people and walking disabilities using a wearable device (WD) and KINECT to detect warning signs of falls. On the basis of the results of experiments, we decided to adopt the angle velocity at the terminal stance to initial swing, angle between a foot and floor at the terminal swing, and average FS to detect warning signs of falls. We also developed a gait monitoring device comprised of a smartphone application and a pair of shoes on which WDs were mounted, and proposed using the side and top view formats to present data measured by KINECT. Proposed warning signs calculated for walking disabilities using developed device and KINECT showed clearly difference from healthy people. The proposed presentation format also made clear difference between them. We plan to develop a system that measures effect of rehabilitation quantitatively, and a warning system for fall prevention using proposed gait monitoring device and presentation formats.

References

[1] S. Yoshida, and Y. Murata, "Proposal of Gait Monitoring Shoes Based on Comparative Analysis of Walking Gait Cycle between Normal People and Stroke Patients," IARIA, eTELEMED 2017, pp. 56-61, 2017.

[2] World Population Ageing: 1950-2050, Population Division, Department of Economic and Social Affairs, United Nations, http://www.un.org/esa/population/publications/worldageing19502050/ [retrieved: November, 2017].

[3] W. P. Berg, H. M. Alessio, E. M. Mills, and C. Tong, "Circumstances and consequences of falls in independent community-dwelling older adults," Age Ageing, Vol. 26, pp. 261–268, 1997.

[4] World Health Organization Regional Office for Europe, "What are the main risk factors for falls amongst older people and what are the most effective interventions to prevent these falls?,"

http://www.euro.who.int/en/data-and-evidence/evidence-informed-policy-making/publications/pre2009/what-are-the-main-risk-factors-for-falls-amongst-older-people-and-what-are-the-most-effective-interventions-to-prevent-these-falls [retrieved: November, 2017]

[5] F. A. Batchelor, S. M. Mackintosh, C. M. Said, and K. D. Hill, "Falls after stroke," Int J Stroke, Vol. 7, pp. 482-90, 2012.

[6] G. S. Verheyden, V. Weerdesteyn, R. M. Pickering, D. Kunkel, S. Lennon, A. C. Geurts, and A. Ashburn, "Interventions for preventing falls in people after stroke," Cochrane Database Syst Rev. (5), CD008728, 2013.

[7] J. A. Tutuarima, J. H. P. van der Meulen, R. J. de Haan, A. van Straten, and M. Limburg, "Risk Factors for Falls of Hospitalized Stroke Patients," Stroke, Vol. 28, pp. 297-301, 1997.

[8] R. Teasell, M. McRae, N. Foley, and A. Bhardwaj, "The Incidence and Consequences of Falls in Stroke Patients During Inpatient Rehabilitation: Factors Associated With High Risk," Arch Phys Med Rehabil Vol. 83, 2002.

[9] R. K. Begg, O. Tirosh, C. M. Said, W. A. Sparrow, N. Steinberg, P. Levinger, and M. P. Galea, "Gait training with real-time augmented toe-ground clearance information decreases tripping risk in older adults and a person with chronic stroke," Frontiers in Human Neuroscience, Vol. 8, Article 243, 2014.

[10] A. J. Blake, K. Morgan, M. J. Bendall, H. Dallosso, S. B. Ebrahim, T. H. Arie, P. H. Fentem, and E. J. Bassey, "Falls by elderly people at home: prevalence and associated factors," Age Ageing; Vol. 17, pp. 365-72, 1988.

[11] Meet Kinect for Windows,

https://developer.microsoft.com/en-us/windows/kinect, [retrieved: November, 2017].

[12] W. Tao, T. Liu, R. Zheng, and H. Feng, "Gait Analysis Using Wearable Sensors," Sensors, 12(2), pp. 2255-2283, 2012.

[13] S. J. M. Bamberg, A. Y. Benbasat, D. M. Scarborough, D. E. Krebs, and J. A. Paradiso, "Gait Analysis Using a Shoe-Integrated Wireless Sensor System," IEEE Transaction on Information technology in Biomedice, Vol. 12, No. 4, 2008.

[14] F. Dadashi, B. Mariani, S. Rochat, C. J. Büla, B. Santos-Eggimann, and K. Aminian, "Gait and Foot Clearance Parameters Obtained Using Shoe-Worn Inertial Sensors in a Large-Population Sample of Older Adults," Sensors 2014, Vol. 14, pp. 443-457, 2014.

[15] M. Benoussaad, B. Sijobert, K. Mombaur, and C. A. Coste, "Robust Foot Clearance Estimation Based on the Integration of Foot-Mounted IMU Acceleration Data," Sensors 2016.

[16] VICON,

http://www.vicon.com/, [retrieved: November, 2017].

[17] S. Obdrzalek et al., "Accuracy and Robustness of Kinect Pose Estimation in the Context of Coaching of Elderly Population," IEEE, Engineering in Medicine and Biology Society (EMBC 2012), 2012 Annual International Conference of the IEEE, 2012.

[18] K. Suzuki, Y. Murata, K. Yoshida, "Remote Rehabilitation System for Cerebrovascular Patients Combined with Video Call Center," IARIA, International Journal on Advances in Life Sciences, vol. 7 no 3 & 4, pp. 132 – 144, 2015

The Wize Sniffer Knows What You Did: Prevent Cardio-Metabolic Risk by Analyzing Your Breath

Danila Germanese
Institute of Information Science and Technology
ISTI-CNR
Via G. Moruzzi 1, Pisa, Italy
Email: danila.germanese@isti.cnr.it

Mario D'Acunto
Institute of Biophysics
IBF-CNR
Via G. Moruzzi 1, Pisa, Italy
Email: mario.dacunto@ibf.cnr.it

Marco Righi and Massimo Magrini and Ovidio Salvetti
Institute of Information Science and Technology
ISTI-CNR
Via G. Moruzzi 1, Pisa, Italy
Email: marco.righi@isti.cnr.it
Email: massimo.magrini@isti.cnr.it
Email: ovidio.salvetti@isti.cnr.it

***Abstract*—Its un-obtrusiveness and its inherent safety make breath analysis a very promising technique in healthcare diagnostics. On one hand, it enables the monitoring of biochemical processes: the volatile organic compounds (VOCs) from the metabolic processes are generated within the body, travel via the blood, participate to the alveolar exchanges and appear in exhaled breath; on the other hand, breath is easily and non-invasively accessible. Nevertheless, despite its great potential, breath analysis is not widely used in clinical practice: the high costs for standard analytical instrumentation (i.e., gas chromatograph-mass spectrometer), the need for specialized personnel able to read the results and the lack of standardized protocols to collect breath samples, set limits to its exploitation. Here, we describe the Wize Sniffer (WS), a portable device based on low cost technology, able to collect and analyze in real time the composition of the breath. In particular, by means of the WS, the user can evaluate his/her own cardio-metabolic risk score by self-monitoring the composition of the breath. Indeed, the presented device is able to detect, in real time, all those VOCs related to the noxious habits for cardio-metabolic risk. Nonetheless, the modular configuration of the WS, makes it usable also for other applications by changing the type of the gas sensors according to the molecules to be detected.**

***Keywords*–Breath analysis; E-noses; Gas sensors; Self-monitoring; Signal processing; Bio-signals; Medical device; Cardio-metabolic risk prevention.**

I. Introduction

Since the time of Hippocrates, classical medicine has used the sense of smell as an indicator of human diseases [1], [2]: the fruity-smelling breath underlined the presence of diabetes; the stale beer-like odor of the skin was typical of the persons with tuberculosis; the butcher's-like smell of the skin suggested yellow fever, etc. Therefore, early medical practitioners recognized that the presence of human diseases changed the odors released from the body and breath.

It was in 1784 when, for the first time, Lavoisier and Laplace identified the presence of carbon dioxide in human exhaled breath. However, it is commonly recognized that the modern breath analysis started in 1971, when Linus Pauling demonstrated that breath is a mixture of more than 200 volatile molecules at the levels of part per million (ppm), part per billion (ppb), or lower [3].

Breath is the product of the composition of inspiratory air, molecules deriving from ingested food and beverages or from dermal adsorption (exogenous molecules), and all the volatile substances in the blood, which are produced endogenously as part of our normal (or disease-related) metabolism and participate to alveolar exchanges according to their types, concentrations, volatilities and rates of diffusion. In addition, also cells in the mouth, upper airways, and gastro-intestinal tract contribute volatile molecules to the breath. Human exhaled breath is composed of nitrogen (75%), oxygen (13%), water vapor (6%), carbon dioxide (5%). The remaining 1% is composed of a series of volatile organic compounds (VOCs) that are peculiar for each individual. As a consequence, it is correct to think that every one of us has his/her own *breath-print*, which can tell a lot about the state of health.

Breath analysis, for its un-obtrusiveness and its inherent safety, may play a very important role in health care diagnostics. It may be used to detect disease, monitor disease progression, or monitor a therapy. Indeed, many studies aimed to correlate breath VOCs to various diseases such as diabetes, lung cancer, gastrointestinal diseases, etc. [4], [5], [6]. Exhaled pentane and ethane were investigated as lipid peroxygenation product in case of oxidative stress [7]; isoprene (the major hydrocarbon present in human breath) was suggested to be linked with cholesterol synthesis [8] and cardiac output [9]; breath ammonia may be a useful biomarkers both for the evaluation of clinical treatments in case of renal diseases [10], [11] and for monitoring the level of severity in case of liver diseases [12]; increased levels of breath carbon monoxide may be due to airway inflammation in asthma and in chronic obstructive pulmonary disease (COPD)[13].

Comparing with other traditional methods such as blood test, breath analysis is non-invasive, real-time, and harmless to not only the subjects but also the personnel who collects the samples. Nonetheless, despite its great potential, only few breath tests (among which: carbon monoxide test for neonatal jaundice, ethanol test for drunk drivers, hydrogen test for the evaluation of gastrointestinal transit time, for the monitoring of intestinal bacterial overgrowth and for the assessing of h.pylori infection, nitric oxide test for the evaluation of asthmatic disease) are commonly used in clinical practice nowadays. In [14] T.H. Risby and S.F. Solga give a fair view of current status of breath analysis and try to explain the reasons why it has not gained a wider use yet. One of such reasons is related to the high costs of the specific, standard instrumentation for gas analysis (i.e., gas chromatograph, mass spectrometer) and the need of expert personnel to perform the analysis, which also are very time consuming [10].

In recent years, the idea of exploiting e-noses also for clinical applications has gained the attention of the scientific community [15]. Being quicker than a gas chromatograph, as they are able to follow the trend in time of breath molecules, in many studies they have been employed in different fields of medicine: in oncology, for instance, to identify lung cancer-related breathprint [16], in infectiology [17], in respiratory medicine to evaluate asthma [13] or to discriminate between healthy subjects and patients suffering from chronic obstructive pulmonary disease (COPD) [18]. Nevertheless, the majority of such e-noses exploit very expensive technology [19], [20] or requires complex circuitry [21], [22].

By developing the Wize Sniffer (WS) [1], [23], described in this paper, we aimed to overcome this limitations:

- the WS is a portable, real-time device, which might be used not only in laboratory settings, but also in doctor's office, or in home environment;
- it is very easy to use, also for non-specialized personnel, thus allowing the self-monitoring of own health state. In addition, it is programmed to send breath analysis results also to a remote care center;
- it is entirely based on low cost technology: the employed gas sensors are commercial, semiconductor-based and easily embeddable in the circuitry; breath signals are analyzed by a widely employed open source controller: Arduino Mega2560.

The WS was developed in the framework of SEMEOTICONS European Project [24], which aimed to develop the Wize Mirror, (WM) a multi-sensory platform having the appearance of a mirror. The WM, by means of a series of depth sensors and multispectral cameras, is able to assess individual's well-being state by detecting in his/her face all those signs related to cardio-metabolic risk [25], [26]. The WS was designed to be a WM's tool, in order to detect in human breath the molecules related to the noxious habits for cardio-metabolic risk: alcohol intake, wrong diet, smoke. Not only: we aimed to develop a device whith a modular core, and which could be also used for broader applications [27], [1].

In the paper, Section II lists the molecules detected by the WS; Section III describes the devices general hardware/software architecture; Section IV explains the WS functionality tests and the experimental results, later discussed in Section V.

II. THE DETECTED VOCs

Our aim was to develop a device, which could help the user to monitor his/her noxious habits for cardio-metabolic risk simply by analysing the breath composition. Therefore, within the WS, an array of semiconductor-based gas sensors is able to detect the following molecules:

- Carbon monoxide (CO): in human body, it is naturally produced by the action of heme oxygenase on heme when the macrophages of the spleen remove old and damaged erythrocytes from the circulation. Also, it is the major compound in cigarette smoke. An increase of CO in blood is very dangerous, as it leads hemoglobin to carry less oxygen through the vessels, because CO usurps the space in hemoglobin that normally carries oxygen, forming carboxyhemoglobin [28]. It also increases the amount of cholesterol that is deposited into the arteries. CO normal levels in exhaled breath are 2-3.5ppm; increasing levels can be detected in smokers (13.8 - 29ppm);
- Ethanol (C_2H_6O): endogenous ethanol levels are 0-3.9ppm (mean 0.62ppm), normally lower than the ones found in subjects' breath after alcoholic drinks ingestion. However, moderate ethanol consumption, in healthy subjects, reduces stress and increases feelings of happiness and wellbeing, and may reduce the risk of coronary heart disease. Heavy consumption of alcohol, instead, causes addiction and leads to an accumulation of free radicals into the cells, causing oxidative stress [29].

The WS can also provide useful information about metabolism, carbohydrates adsorption and vascular status by detecting:

- Oxygen and carbon dioxide (O_2 and CO_2): exhaled air has a decreased amount of oxygen and an increased amount of carbon dioxide. These amounts show how much O_2 is retained within the body for use by the cells and how much CO_2 is produced as a by-product of cellular metabolism. Exhaled O_2 amount is about 13.6%-16%. Mean CO_2 concentration in exhaled breath is about 4% (= 40000ppm) [30]. Individual's breathing rate influences the level of CO_2 in blood and, as a consequence, in exhaled gas. Breathing that is too slow causes respiratory acidosis (that results in an increase of CO_2 partial pressure in blood, which may cause hypertension), while breathing that is too rapid causes a decrease in CO_2 in blood that leads to hyperventilation and respiratory alkalosis;
- Hydrogen (H_2): it is related to the carbohydrates breakdown in the intestine and in the oral cavity by anaerobic bacteria [31]. Breath hydrogen levels vary within a day and from day to day; fasting levels range between 0.3 and 34.1ppm (mean 9.1ppm). However, it may vary also among individuals, especially in case of lactose intolerance and celiac diseases;
- Hydrogen sulfide (H_2S): in healthy subjects, concentrations of volatile sulfur-containing compounds in blood are very low. The body uses sulfur compounds in order to neutralize the action of free radicals [32]. Among the sulfur-containing volatile molecules, hydrogen sulfide is considered as a vascular relaxant agent, as it has a therapeutic effect in various cardio-vascular diseases.

III. WHITHIN THE WS: HARDWARE AND SOFTWARE

Here, we describe the hardware and software platforms of the WS.

A. Hardware

In designing our device, we took into account an important issue: the greater demands on improvements in effectiveness, smartness and lower costs of biomedical instruments used in daily healthcare applications [33], resulted from increasing limitations of healthcare financial resources as a consequence of budgetary cuts or constraints.

For this reason, we aimed to design a device entirely based on low-cost technology. In Figures 1, 2 and 3, WS' hardware is shown.

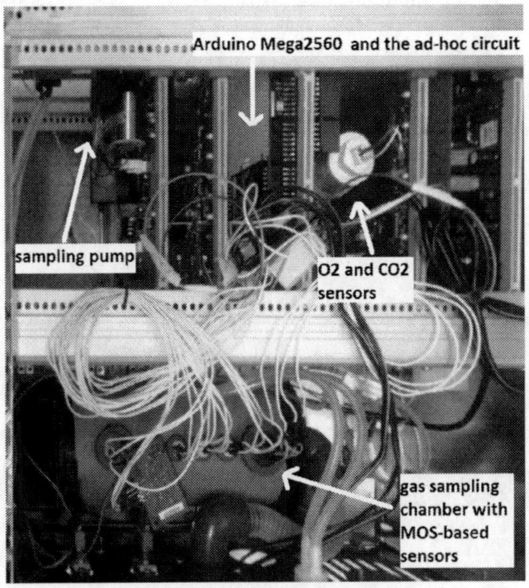

Figure 1. Wize Sniffer's hardware, internal configuration

Figure 2. Wize Sniffer's gas sampling chamber detail

The exhaled gases flow into a corrugated tube, made of polyvinyl chloride (PVC), and reach the gas sampling chamber. A heat and moisture (HME) filter, made of hygroscopic material, absorbs the water vapor present in the exhaled breath: as it

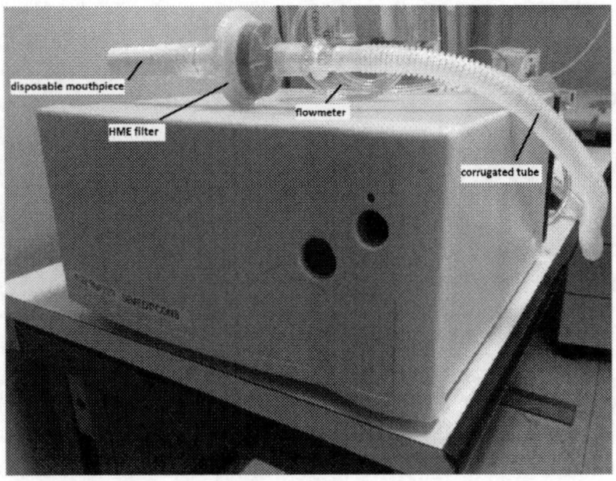

Figure 3. Wize Sniffer's hardware, external configuration. Its dimensions are: 30x30x14cm

will be described later, the employed gas sensors'conductivity response is strongly affected by humidity. Integrating such a filter allows for reducing the humidity of -30% and also for holding users' oral bacteria. A PNT Flow-Ree flowmeter allows for monitoring the user's flow rate and for calculating the exhaled gas volume. The core of the WS is the signal measurement module, that is the sensor array, composed of six semiconductor-based gas sensors, placed within the gas sampling box. The latter was made up of acrylonitrile-butadiene-styrene (ABS) and Delrin, which are two materials that do not interfere with sensors's sensitivity, and its capacity is 600ml according to the tidal volume [30]. Within the gas sampling box, also a sensor for temperature and humidity (Sensirion SHT11) is placed. In addition, a sampling pump injects, at a fixed rate (120ml/sec), the sampled exhaled gas to other two sensors, which have faster response time and work in *flowing-regime*. They detect oxygen and carbon dioxide and are respectively based on an electrochemical cell and an infrared source. Sensors' raw output are pre-processed and stabilised by a signal conditioning module: a series of voltage buffer amplifiers (LM124-N, Texas Instrument) transfers sensors' signals from the measurement module to a widely employed open source controller: an Arduino Mega2560 with Ethernet module. Finally, in order to facilitate sensors' recovery time, a flushing pump was integrated on one side of sampling chamber. After each breath test, it can be switched on in order to "purge" the chamber with ambient air and recovery sensors' baseline.

In Table I, the employed gas sensors are listed. Our aim was to find a trade-off between good sensitivity, low cost and small dimension. As we mentioned in the previous Section, the WS was developed to detect a set of molecules related to those noxious habits for cardio-metabolic risk; nevertheless, our aim was to design a modular sensor platform in order to detect a broader set of molecules, simply by changing the sensors according to the VOCs to be identified. As a consequence, the sensors' ease of integration in the circuitry was another requirement we needed.

Optical, carbon nano-fiber (CNF), quartz crystal microbalance (QCM), metal oxide semiconductors (MOS), conducting polymers (CP), and surface acoustic wave (SAW), are the most common gas sensor types employed in e-noses [15], [34].

On one hand, optical gas sensors, as well as quartz crystal microbalance (QCM)-based gas sensors and surface acoustic wave (SAW)-based gas sensors are very sensitive; on the other hand, they are expensive (especially in the case of optical gas sensors) and need complex circuitry (in the case of QCM and SAW gas sensors). Also carbon nano-ber (CNF) based gas sensors are very expensive, especially for their manufacturing. We chose metal oxide semiconductor (MOS)- based gas sensors: they show long term stability and reproducibility of gas response [35], great metallurgical and chemical stability of the sensing material [35], high sensitivity towards target gases, short reaction and recovery time, easy calibration. In comparison to other types of gas sensors, MOS-based gas sensors' availability, small dimensions, compactness and low cost make them the most widely used gas sensors [36].

Whithin the gas sampling chamber, six Taguchi semiconductor-based gas sensors, manufactured by Figaro Engineering [37] (costs: 25-40 Euro), were integrated.

TABLE I. SENSORS INTEGRATED IN THE WS SENSOR PLATFORM

Detected molecule	*Sensor*	*Best detection range*
Carbon monoxide	TGS2442	50-1000ppm
	MQ7	20-200ppm
	TGS2620	50-5000ppm
Ethanol	TGS2602	1-10ppm
	TGS2620	50-5000ppm
Carbon dioxide	TGS4161	0-40000ppm
Oxygen	MOX20	0-16%
Hydrogen sulfide	TGS2602	1-10ppm
Hydrogen	TGS821	10-5000ppm
	TGS2602	1-10ppm
	TGS2620	50-5000ppm
	MQ7	20-200ppm

Unfortunately, humidity and cross-sensitivity strongly affect the behavior of MOS-based gas sensors [35], as shown in Figure 4. The water vapour undergoes dissociative adsorption on metal oxide surfaces and the resultant ions are adsorbed on the metal oxide surface, impeding, in many cases, the response of sensor by lowering the sensitivity of the sensing element [38], [39]. However, a distinction should be done between n-type and p-type-based sensing materials [40]. As shown in Figure 4, when humidity icreases, resistance of the n-type-based film decreases, resulting in a rise in voltage output.

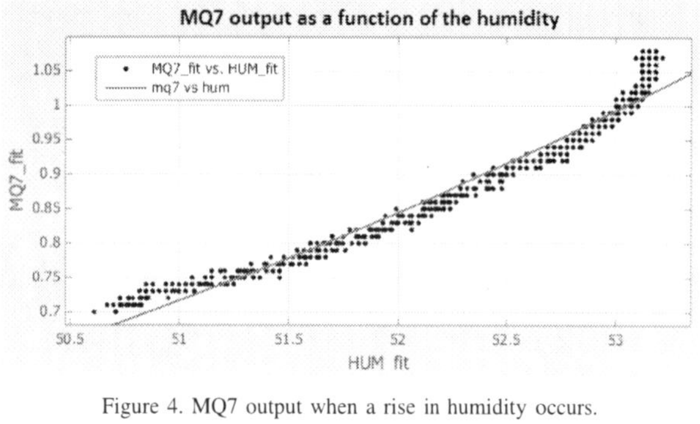

Figure 4. MQ7 output when a rise in humidity occurs.

In our case, humidity plays a very crucial role, as we deal with human breath. Therefore, we deemed it necessary to take steps to manage this factor and optimize its effects. First, as previously described, we put a humidity filter, made of hygroscopic material, in order to absorb the majority of the water vapor present in exhaled breath and reduce the humidity in the gas sampling chamber from a 90% to a 70-60%. In addition, we integrated, into the gas sampling chamber, also a temperature and humidity sensor (Sensirion SHT11) in order to monitor these two parameters. Indeed, we noted that the temperature inside the chamber remains almost constant before, during and after each breath test. On the contrary, relative humidity shows a variation of about 35% while the subject is performing a breath test (Figure 5). For this purpose, we i) calculated sensors' drift due to variations in humidity; ii) investigated sensors' sensitivity in precise measurement conditions [1] (30 °C±7%, 70%RH±5%, that are the ones that occur in the sampling box during a breath test, as shown in Figure 5.

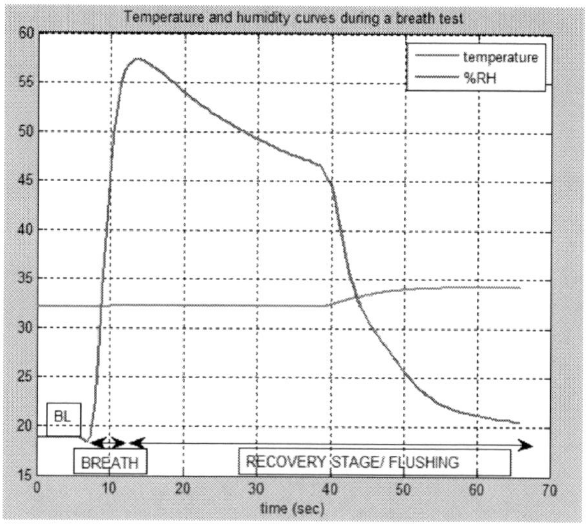

Figure 5. Temperature and relative humidity in the gas sampling box when a breath analysis is performed

The relationship between humidity and sensors' output V_{out} (as can be observed in Figure 4) generally can be modeled by means of a power law (eq. (1)), as reported also by Ho Sohn and coworkers [41]:

$$V_{out} = f(hum) = a * (hum^b) + c \tag{1}$$

where a, b and c are constant and specific for each TGS sensor. Calculating sensors' humidity drift is useful to potentially compensate it during the data processing. We considered the entire range of humidity variation (for instance, 50%-55%RH in the case of MQ7) and then we calculated the slope of the curves. Based on the slope, drift coefficient S_d was assessed for each sensor (see Table II) as the decrease in sensors output ΔV (Volt) per unit decrease in humidity, Δhum as given in eq. 2:

$$S_d = \frac{\Delta V}{\Delta hum} \tag{2}$$

By keeping the humidity constant, sensors' output depends on the gas concentration only. For this purpose, we investigated sensors' behavior in response to a well-known gas concentrations at a fixed humidity and temperature conditions. The used experimental set-up was the one reported in Figure 6.

TABLE II. SENSORS DRIFT DUE TO HUMIDITY

Sensor	ΔV / Δhum (mV)
MQ7	296
TGS2620	60
TGS2602	82
TGS821	120
TGS2444	84

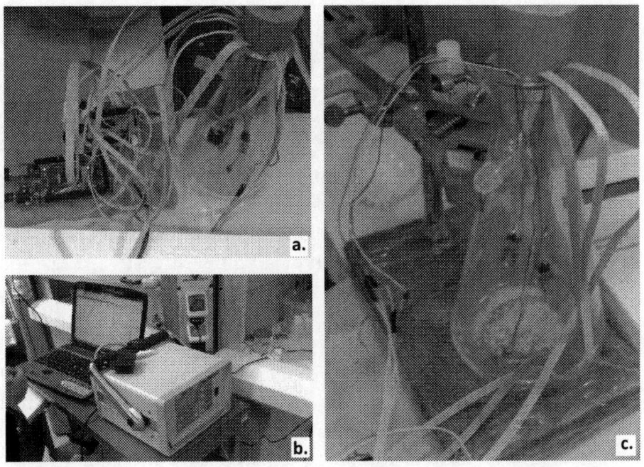

Figure 6. Experimental set-up. a) The gas sensors' raw output are read by an Arduino Mega2560 connected via USB to a personal computer; b) the signals are displayed in real time on the computer screen; c) the gas sensors into the vial, where is placed a saturated solution of $NaCl$ on the bottom.

The gas sensors were put into a vial, where the humidity was kept at 70%RH$\pm$5% by means of a saturated solution of $NaCl$ placed on the bottom. Measurements were performed only after the sensors were operated at a fixed temperature for several hours (at least 2 hours for warm-up).

Then, we injected well-known gases concentration and registered sensors' output. The raw sensors output were read by an Arduino Mega2560 connected via serial port to a personal computer. The experimental data were displayed in real time on the computer screen and stored as text files for later processing. For example, in Figure 7, we can see TGS2620 output when well-known concentrations of carbon monoxide, ethanol and hydrogen were separately injected into the vial.

In general, the relationship between sensors' output V_{out} and each gas contribution can be modeled by means of eq. (3).

$$V_{out} = f([gas]) = a * ([gas]^b) + c \tag{3}$$

Where c is a constant, b is the constant power-law exponent and a can be considered as the sensor's sensitivity coefficient. These parameters are specific for each TGS sensor. We also found that the used gas sensors were sensitive to concentration lower than their best detection range (reported in Table I).

Nevertheless, when a breath analysis is performed, a mixture of gases spreads into the gas sampling box and chemically interacts with the sensors. In this case, the phenomenon known as cross sensitivity makes the semiconductor sensors non-selective. In addition, the detection threshold (that is, the minimum concentration of gas necessary to a meaningful change in sensors' conductivity) depends not only on absolute

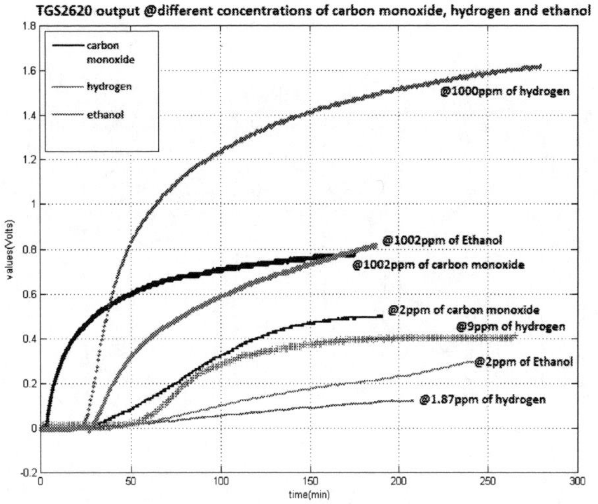

Figure 7. TGS2620 output when well-known concentrations of CO, H_2 and C_2H_6O were separately injected into the vial.

sensitivity to that particular gas but also on concentrations of the other gases, which partially mask the response to the gas of interest, as shown in eq. (4) reported by P. Clifford and coworkers [35] (the equation refers to the case of only two mixed gases, as an example).

$$(\frac{R_j}{R_{0j}})^{-\frac{1}{\beta_j}} = \frac{(1 + \sum K_j * [G_{1j}]^{n_{1j}} * K_j * [G_{2j}]^{n_{2j}})}{[O_2]} \tag{4}$$

Where $\frac{R}{R_0}$ is the j-th sensor's variation in resistance, β is the power law exponent, specific for each j-th sensor, $[G_1]$ and $[G_2]$ represent the concentrations of the two mixed gases, n is an integer or fractional integer power, specific for each j-th sensor, K can be considered as the j-th sensor's sensitivity coefficient and $[O_2]$ is the oxygen partial pressure. In some cases, for some terms of the summation, there is only one term per gas, for oters there is a product of several.

We also tried to investigate the cross sensitivity of our sensors. In Figure 8, we can see TGS2620 response when well-known mixed concentrations of the three gases (carbon monoxide, ethanol and hydrogen) were injected into the vial at the same time.

In this way, how the different VOCs add together and influence gas sensors output can be understood. The single gas contribution can be modeled by a power law similar to eq. (3), but each of them has its "weight" on the overall output, as shown in eq. (4). However, because of the multitude of factors involved, understanding the interaction mechanism behind the MOS-based gas sensors' sensing property in general remains an open issue [40].

Finally, in Figure 9, the WS performing a breath test is shown.

B. The Software

Given its un-obtrusiveness and its safety, breath analysis may be used as a daily monitoring analysis tool. To fully

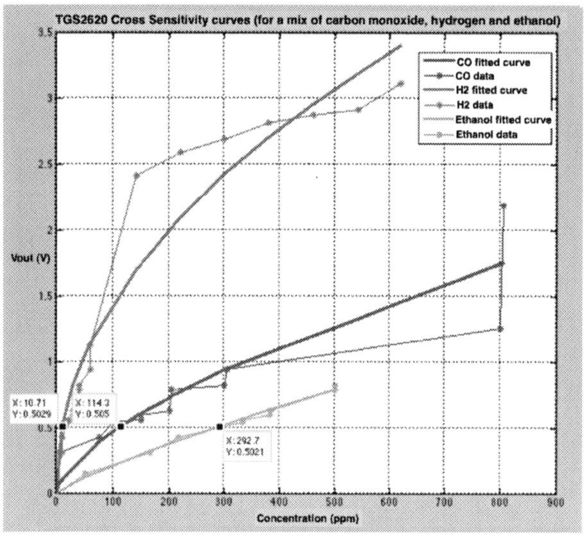

Figure 8. TGS2620 output when well-known mixed concentrations of CO, H_2 and C_2H_6O were contemporarily injected into the vial.

Figure 9. The WS while performing a breath test.

exploited its potential, its application must take place not only in laboratory settings, but also in the clinics, in doctors' offices, and at home. Our aim was to develop a device, which could be used in home environment and which could be able to send breath analysis results also to a remote personal computer (for instance, to the one of the own family doctor).

Arduino Mega2560 with Ethernet board samples sensors signals every 250msec, saves raw vector data and extracts the maximum value of raw breath curve. Several parameter and features can be derived from breath curves [42] to fully characterized them. We chose to calculate the value at curve plateau as it better describes the chemical balance between sensorssensing element and target gases. Such data are then processed and analyzed, as described in Section IV. In order to send breath analysis results also to a remote personal computer, we implemented a client-server architecture. It means that, after performing a test and processing the results, the device, by means of an internet connection and a communication protocol, can send the results to the physician, for instance. Arduino is programmed to execute a daemon on port 23. By implementing a Telnet server, it waits for a command line from the remote personal computer and then can provide the data. A measure is valid if the users exhaled volume equals at least the one of gas sampling box (600ml).

IV. WS VALIDATION

The aim of the validation was to assess if the WS was able to monitor and evaluate the individuals noxious habits for cardio-metabolic risk (smoke and alcohol intake in particular). For this purpose, as described in [1], the WS underwent a clinical validation in three research centers: CNR in Pisa and Milan, CRNH (Centre de Recherche en Nutrition Humaine) in Lyon. The campaign involved 77 volunteers, male and female, between 30-65 years of age, with different habits and lifestyles. People answered Audit and Fagerstrom tests, which respectively assessed their alcohol and smoke dependence, and other questionnaires about lifestyle in general.

Exhaled breath composition is strongly inuenced by breath sampling method [43], as well as by breath flow rate [44], posture [45], ambient air [46], lung volume [47]. In our case, also factors such as BMI, sex, age, subjects lifestyle may inuence the breath composition: for example, alcohol disposal in men is different than the one in women, and, in addition, it may depend on body mass index (BMI)[48], as well. Therefore, breath composition may exhibit not only a strong inter-variability (among different subjects), but also a marked intra-variability (relative to the same subject). Moreover, standard protocols for breath sampling do not exist. The definition of precise guidelines to collect breath sample would be useful also to avoid such factors that influence the breath composition.

As a consequence, we considered, on one hand, all these issues about human breath variability and influencing factors, and, on the other hand, the methodological issues about breath sampling [43]. Since our interest was focused on both endogenous and exogenous biomarkers, we drafted a protocol which exploited the mixed expiratory air sampling. Such method of sampling entails collecting total breath, including the air contained in the upper airways, which involves volatile compounds that do not participate to alveolar exchanges (dead space air) [46]. The subjects took a deep breath in, held the breath for 10s, and then exhaled once into the corrugated tube trying to keep the expiratory flow constant and to completely empty their lungs. However, mixed expired air consists of dead space, transition phase and alveolar phase. The dead space and the transition phase contain breath compounds from the upper airways, whereas the alveolar phase contains the VOCs resulting from alveolar exchanges, which better represent the individuals metabolic conditions. As a consequence, a controlled identification of the respiratory phases was performed, by monitoring the curve of exhaled carbon dioxide (i.e., capnogram) and discriminating between dead space air fraction and end-tidal exhaled air. In addition, the applied manoeuvres such as breath holding [49], high exhaled volume, lower exhalation flow rate [50], [51], [52] and single exhalation [53] lead to an incrase in alveolar VOCs concentrations in breath samples, thus reducing the impact of the previously listed influencing factors (especially ambient air).

The study was approved by the Ethical Committee of the Azienda Ospedaliera Universitaria Pisana, protocol n.213/2014 approved on September 25th, 2014; all patients provided a signed informed consent before enrollment.

As mentioned before, MOS-based gas sensors are not selective, thus impeding to calculate the exact molecules' concentrations. Multivariate methods of pattern recognition techniques usually allow for overcoming this problem. Pattern recognition exploits sensors' cross-correlation and helps to extract qualitative information contained in sensors' outputs ensemble. Therefore, first Principal Component Analysis (PCA) was performed, in order to provide a representation of the data in a space of dimensions lower than the original sensors' multidimensional space. From a visual, exploratory analysis of the data, the presence of clusters (see Figure 10) was observed. Then, a Knearest neighbor (KNN) classification algorithm, previously trained with the data collected during another acquisition campaign, was adopted to classify the subjects according to their habits: Healthy (that means, not in danger of cardio-metabolic diseases), Light Smoker, Heavy Smoker, Social Drinker, Heavy Drinker, LsSd (Light smoker and Social drinker), LsHd (Light smoker and Heavy drinker), HsSd (Heavy smoker and Social drinker), HsHd (Heavy smoker and Heavy drinker).

The outcomes of the Audit and Fagerstrom questionnaires were our ground truth. The KNN classifier was able to correctly classify in 89,61% of cases. Errors were probably due to TGS2602 and TGS2620 cross-sensitivity for hydrogen. In fact, for instance, three no-risk subjects were classified as social drinker probably because of the high presence of hydrogen in their breath, which caused a rise in these sensors' voltage output.

Then, the number of volunteers increased up to 169 subjects. They were classified by clinicians into "low risk population", "medium risk population", "high risk population" and "very high risk population", on the base of their Risk Score (RS), that is, the sum of the scores relative to Audit (AS), Fagerstrom (FS) and lifestyle questionnaires, our ground truth also in this case. Given the significant number of subjects, we tried to implement a simple model, which was able to predict subjects RS on the base of breath data. First, sensors' raw data were zero-centered and normalized, thus putting in evidence their qualitative aspects. Then, also in this case, the principal components were extracted and the PC scores were plotted against the subjects RS, as shown in Figure 11.

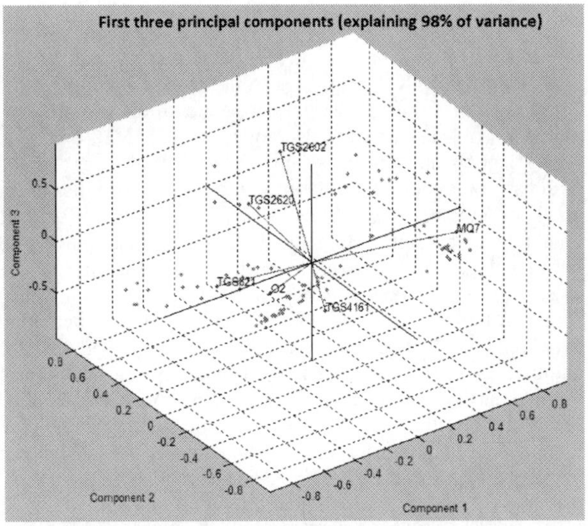

Figure 10. First three principal components. The presence of several clusters can be observed.

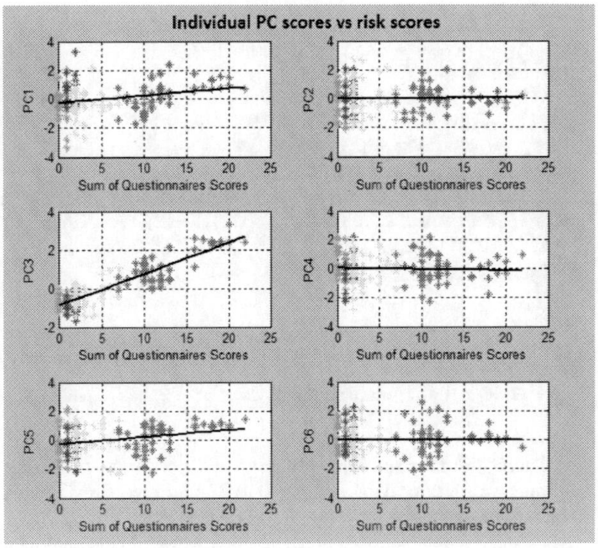

Figure 11. PC scores against subjects risk scores arranged in ascending order.

As can be deduced from the colours (green points derive from no-risk subjects, the blue ones from low-risk subjects, the yellow ones from medium-risk subjects, the red ones from high-risk subjects, the magenta ones from very high-risk subjects), subjects' RS were arranged in ascending order. Except for PC3, from an exploratory analysis, we saw that the PC scores did not have a sharp increasing or decreasing linear trend with respect to RS, thus not having enough information to contribute to any prediction model. Such result matched the one reported in [54]. Being inspired by this study, we also implemented an Independent Component Analysis (ICA) on our data. ICA is a high-order transformation method for data representation, which extracts independent component from the data set. If, on one hand, PCA exploits the real sensors cross-correlation, ICA originates from the assumption that the data has a non-Gaussian distribution, which often is a property of the gas sensors array measurement data [55]. In our case, breath signals and the environmental ones (noise) get mixed with each other before the chemical interaction with the sensor array. In addition, due to sensors' cross-sensitivity effect, the conductivity response depends on a linear combination of individual gas terms, where the effects of one gas can be masked by the combined effects of others. In addition to this "competition" among gases, there is an associative interaction by which the effects of one gas are enhanced by the presence of another. As a consequence, each sensors output is the result of a combination of different gaseous contributions. We applied FastICA algorithm to our data set, and plotted individual independent components (IC) against subjects' RS. As shown in Figure 12, in this case sharper linear trends emerged.

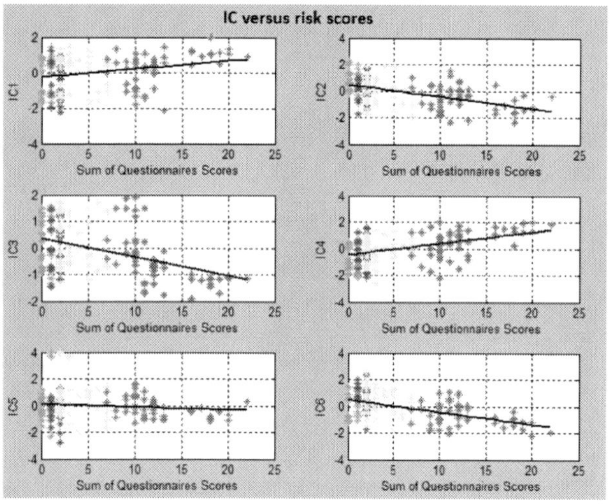

Figure 12. IC scores against subjects risk scores arranged in ascending order.

Then, the data set was split into two data-sets (train data set and validation data set) to build the prediction model, which was developed by means of the Matlab LinearModel Tool. Indeed, by using the independent components, a linear regression model was built to establish a relationship between the volunteers' RS and the breath data pre-processed by ICA. Then, such model was validated by using the validation data set. In Figure 13 we can see that the correlation coefficient (r) between actual and estimated risk scores was 0.8976.

V. CONCLUSION

The field of breath analysis is as old as the one of medicine. Since the time of Hippocrates, classical medicine has exploited the sense of smell to identify human diseases.

For its un-obtrusiveness, and its inherent safety, breath analysis may be a very useful tool in health care diagnostics: to detect disease, or to monitor disease progression, or to evaluate the success of a therapy. With respect to other traditional methods such as blood test, breath analysis is non-invasive, real-time, and harmless to not only the subjects but also the personnel who collect the samples. Nevertheless, to fully exploit such potential, breath analysis should be performed

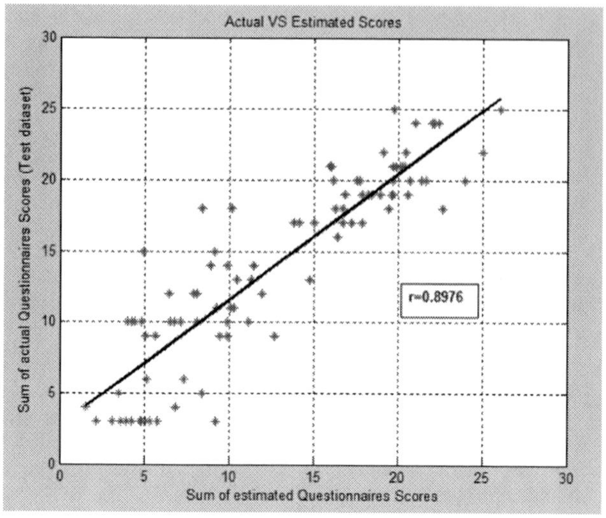

Figure 13. Actual risk scores versus predicted ones

by using a simple, portable, easy to use instrumentation. In addition, an important issue should be considered, that is the greater demands on improvements in lower costs of biomedical instruments used in daily healthcare applications, resulted from increasing limitations of healthcare financial resources as a consequence of budgetary cuts or constraints.

In this paper, we presented the Wize Sniffer, a portable device, able to analyze in real time the composition of human breath, and entirely based on low-cost technology: an Arduino Mega 2560 (a widely employed open source controller) and a semiconductor-based Taguchi gas sensor array.

Although such type of sensors are strongly affected by humidity (which in our case plays a crucial role) with a few, inexpensive arrangements we faced this issue, by integrating a hygroscopic filter behind the mouthpiece, a temperature and humidity sensor within the gas sampling box and by calculating sensors'drift due to humidity. In addition, a robust data processing helped us to face with sensors' cross-sensitivity and human breath inter-variability and intra-variability. When performing a breath analysis by means of non-selective gas sensors, one has to face, on one hand, with an uncertainty of measure which derives from all those factors that affect the gas sensors' behavior; on the other hand, one deals with un uncertainty due to all the physiological conditions that may inuence the breath composition. Pattern recognition algorithms turn out the best way to overcome such problem.

The Wize Sniffer is very simple to use, also by non-specialized personnel: in the presented use case, it provides the user with a very easily interpretable outcome. The WS is able to calculate, by means of a simple regression model, the user's Risk Rcore (with respect to his/her noxious habits) and potentially help him/her to prevent his/her cardiometabolic risk.

However, the user can also send his/her breath analysis results and Risk Score to a remote healthcare center by means of an internet connection: indeeed, Arduino is programmed to implement a simple Telnet Server.

Finally, a strong point of the Wize Sniffer is also its adjustability: thanks to its modular design, the gas sensors can be changed according to the molecules (and then, to the related diseases) to be monitored. Such characteristic allows for using such device in broader applications. Indeed, recently the WS has been used to discriminate and monitor patients with acute liver diseases by evaluating ammonia levels in their breath. The evaluation of WS performances also in the case of cirrhotic patients will be discussed in a future work.

In general, the safety and the un-obtrusiveness of a device for breath analysis, like the presented one, allow for a daily monitoring which, even if without a real diagnostic meaning yet, could represent a pre-screening, useful for an optimal selection of more standard medical analysis. Undoubtedly, further big efforts will be done in order to introduce breath analysis in clinical practice. Scientists and manufacturers should collaborate in order to standardize, on one hand, the architectural principles which e-noses have to be based on, and on the other hand, the procedures for breath sampling, in order to obtain compatible signals and outcomes that may be used and processed by different e-nose systems and shared among physicians all over the world.

ACKNOWLEDGMENT

This work was funded in the framework of the Collaborative European Project SEMEOTICONS (SEMEiotic Oriented Technology for Individuals CardiOmetabolic risk selfassessmeNt and Self- monitoring), grant N. 611516.
Paolo Paradisi, Massimo Martinelli, Massimo Guidi and COSMED s.r.l. are warmly acknowledged for useful support.

REFERENCES

[1] D. Germanese, M. D'Acunto, M. Magrini, M. Righi, and O. Salvetti, "A Low Cost Technology-based Device for Breath Analysis and Self-monitoring," in Proceedings of the Second International Conference on Advances in Signal Image and Video Processing, SIGNAL2017. IARIA, May 2017, pp. 8–13.

[2] J. Pereira et al., "Breath analysis as a potential and non-invasive frontier in disease diagnosis: an overview," Metabolites, vol. 5, 2015, pp. 3–55.

[3] R. Dweik and A. Amann, "Exhaled breath analysis: the new frontier in medical testing," Journal of Breath Research, vol. 2, 2008, p. 030301.

[4] F. DiFrancesco, R. Fuoco, M. Trivella, and A. Ceccarini, "Breath analysis: trends in techniques and clinical applications," Microchemical Journal, vol. 79, 2005, pp. 405–10.

[5] D. Hill and R. Binions, "Breath analysis for medical diagnosis," International Journal on Smart Sensing and Intelligent Systems, vol. 5, 2012, pp. 465–498.

[6] J. Fenske and S. Paulson, "Human breath emission of VOCs," Journal of the Air and Waste Management Association, vol. 49, 1999, pp. 594–598.

[7] M. Phillips, R. Cataneo, J. Greenberg, R. Grodman, and M. Salasar, "Breath markers in oxidative stress in patients with unstable angina," Heart Disease, vol. 5, 2003, pp. 95–99.

[8] R. Salerno-Kennedy and D. Cashmark, "Potential applications of breath isoprene as a biomarker in modern medicine: a concise overview," Wienklin Wochenschr, vol. 17, 2005, pp. 180–186.

[9] F. Pabst, W. Miekisch, P. Fuchs, S. Kischkel, and J. Schubert, "Monitoring of oxidative and metabolic stress during cardiac surgery by means of breath biomarkers: an observational study," Journal of Cardiothoracic Surgery, vol. 2, 2007, p. 9pp.

[10] D. Guo, D. Zhang, N. Li, L. Zhang, and J. Yang, "A novel breath analysis system based on electronic olfaction," IEEE Transaction on Biomedical Engineering, vol. 57, 2010, pp. 2753–2763.

[11] I. Essiet, "Diagnosis of kidney failure by analysis of the concentration of ammonia in exhaled breath," Journal of Emerging Trends in Engineering and Applied Sciences, vol. 6, 2013, pp. 859–862.

[12] R. Adrover et al., "Breath-Ammonia Testing of Healthy Subjects and Patients with Cirrhosis," Digestive Diseases and Sciences, vol. 57, 2012, pp. 189–195.

[13] P. Montuschi, N. Mores, A. Trov, C. Mondino, and P. Barnes, "The electronic nose in respiratory medicine," Respiration, vol. 85, 2013, pp. 72–84.

[14] T. Risby and S. Solga, "Current status of clinical breath analysis," Applied Physics B, vol. 85, 2006, pp. 421–426.

[15] A. Wilson, "Recent progress in the design and clinical development of electronic-nose technologies," Nanobiosensors in Disease Diagnosis, vol. 5, 2016, pp. 15–27.

[16] R. Machado et al., "Detection of lung cancer by sensor array analysis of exhaled breath," American J. of Respiratory and Critical Care Medicine, vol. 171, 2005, pp. 1286–1291.

[17] N. Yusuf et al., "In-vitro diagnosis of single and poly microbial species targeted for diabetic foot infection using e-nose technology," BMC Bionformatics, vol. 16, 2015, pp. 158–169.

[18] O. Sibila et al., "Identification of airway bacterial colonization by an electronic nose in chronic obstructive pulmonary disease," Respiratory Medicine, vol. 108, 2014, pp. 1608–1614.

[19] URL: http://www.bedfont.com/ [accessed: 2017-08-30].

[20] D. Walt et al., "Optical sensor array for odor recognition," Biosensors and Bioelectronics, vol. 13, 1998, pp. 697–699.

[21] F. Lai and H. Huang, "Fabrication of high frequency and low-cost surface-acustic wave lters using near eld phase shift photolitography," Microelectronic Engineering, vol. 83, 2006, pp. 1407–1409.

[22] L. Fan, H. Ge, S. Zhang, H. Zhang, and J. Zhu, "Optimization of sensitivity induced by surface conductivity and adsorbed mass in surface acoustic wave gas sensors," Sensor Act. B-Chem, vol. 161, 2012, pp. 114–123.

[23] M. D'Acunto et al., "Wize sniffer - a new portable device designed for selective olfaction," in Proceedings of the Seventh International Conference on Health Informatics, HEALTHINF2014. SCITEPRESS Digital Library, Mar. 2014, pp. 9–14.

[24] URL: http://www.semeoticons.eu/ [accessed: 2017-08-30].

[25] S. Colantonio et al., "A smart mirror to promote a healthy lifestyle," Biosystems Engineering,Special Issue: Innovations in Medicine and Healthcare, vol. 138, 2015, pp. 33–43.

[26] P. Henriquez et al., "Mirror mirror on the wall... an unobtrusive intelligent multisensory mirror for well-being status self-assessment and visualization," IEEE Transaction on Multimedia, vol. 19, 2017, pp. 1467–1481.

[27] D. Germanese et al., "A low cost, portable device for breath analysis and self-monitoring, the Wize Sniffer," De Gloria A. (eds) Applications in Electronics Pervading Industry, Environment and Society. ApplePies 2016. Lecture Notes in Electrical Engineering, 2017, pp. 51–57.

[28] S. Zevin, S. G. Saunders, S. G. Gourlay, P. Jacob, and N. Benowitz, "Cardiovascular effects of carbon monoxide and cigarette smoking," Journal of the American College of Cardiology, vol. 38, 2001, pp. 1633–1638.

[29] T. Risby, Volatile organic compounds as markers in normal and diseased states. ed. N. Marczin and M. Yacoub, IOS Press, 2002, pp. 113–122.

[30] D. Shier, J. Butler, and R. Lewis, Eds., Holes Human Anatomy and Physiology. McGraw-Hill, 11th Ed, 2007.

[31] D. Thompson et al., "Extra intestinal influences on exhaled breath hydrogen measurements during the investigation of gastrointestinal disease," Gut, vol. 26, 1985, pp. 1349–1352.

[32] W. Miekisch, J. Schubert, and G. Noeldge-Schomburg, "Diagnostic potential of breath analysis- focus on volatile organic compounds," Clinica Chimica Acta, vol. 2004, 347, pp. 25–39.

[33] A. Wilson and M. Baietto, "Applications and advances in electronic-nose technologies," Sensors, vol. 9, 2009, pp. 5099–5148.

[34] K. Arshak, E. Moore, G. Lyons, F. Harris, and S. Clifford, "A review of gas sensors employed in electronic nose applications," Sensor Review, vol. 24, 2004, pp. 181–198.

[35] P. Clifford and D. Tuma, "Characteristics of semiconductor gas sensors I. Steady state gas response," Sens Actuators, vol. 3, 1982, pp. 233–254.

[36] H. Nagle, R. Gutierrez-Osuna, and S. Schiffman, "The How and Why of Electronic Noses," IEEE Spectrum, vol. 1998, 35, pp. 22–31.

[37] URL: http://www.figarosensor.com/ [accessed: 2017-08-30].

[38] C. Wang, L. Yin, L. Zhang, D. Xiang, and R. Gao, "Metal Oxide Gas Sensors: Sensitivity and Influencing Factors," Sensors, vol. 10, 2010, pp. 2088–2106.

[39] Q. Qi et al., "Electrical Response of Sm_2O_3- Doped SnO_2 to C_2H_2 and Effect of Humidity Interference," Sens. and Actuat. B, vol. 134, 2008, pp. 36–42.

[40] P. Shankar, J. Bosco, and B. Rayappan, "Gas sensing mechanism of metal oxides: The role of ambient atmosphere, type of semiconductor and gases - A review," ScienceJet, vol. 4, 2015, pp. 1–18.

[41] J. Ho Sohn, M. Atzeni, L. Zeller, and G. Pioggia, "Characterisation of humidity dependence of a metal oxide semiconductor sensor array using partial least squares," Sens. and Actuat. B, vol. 131, 2008, pp. 230–235.

[42] J. Yan et al., "Electronic nose feature extraction methods: A review," Sensors, vol. 15, 2015, pp. 27 804–27 831.

[43] W. Miekisch, S. Kischkel, A. Sawacki, T. Lieban, M. Mieth, and J. Schubert, "Impact of sampling procedures on the result of breath analysis," Journal of Breath Research, vol. 2, 2008, p. 026007 (7pp).

[44] J. Jones and S. Clarke, "Effect of expiratory flow rate on regional lung emptying," Clin. Sci., vol. 37, 1969, pp. 343–356.

[45] N. Anthonisen, P. Robertson, and W. Ross, "Gravity-dependent sequential emptying of lung regions," Journal of Appl. Physiol., vol. 28, 1970, pp. 589–595.

[46] F. Di Francesco et al., "Implementation of fowlers method for end-tidal air sampling," Journal of Breath Research, vol. 2, 2008, p. 037009 (10pp).

[47] J. Jones, "The effect of pre-inspiratory lung volumes on the result of the single breath O2 test," Respiratory Physiology, vol. 2, 1967, pp. 375–85.

[48] K. Maudens, L. Patteet, A. van Nuijs, C. Van Broekhoven, A. Covaci, and H. Neels, "The influence of the body mass index (BMI) on the volume of distribution of ethanol," Forensic Sci Int., vol. 243, 2014, pp. 74–78.

[49] M. Larstad, K. Toren, B. Bake, and A. Olin, "Determination of ethane, pentane and isoprene in exhaled air effects of breath-holding, flow rate and purified air," Acta Physiol., vol. 189, 2007, p. 8798.

[50] J. Anderson, W. Lamm, and M. Hlastala, "Measuring airway exchange of endogenous acetone using a single-exhalation breathing maneuver," J. of Applied Physiology, vol. 100, 2006, pp. 880–889.

[51] M. Hlastala and J. Anderson, "The impact of breathing pattern and lung size on the alcohol breath test," Ann. Biomed. Eng., vol. 35, 2007, p. 264272.

[52] A. Jones, "Quantitative measurement of the alcohol concentration and the temperature of breath during a prolonged exhalation," Acta Physiol. Scand., vol. 114, 1982, p. 407412.

[53] S. George, A. Babb, and M. Hlastala, "Dynamics of soluble gas exchange in the airways: III. Single exhalation breathing maneuver," J. of Applied Physiology, vol. 75, 1993, p. 243949.

[54] S. Balasubramanian, S. Paniraghi, C. Logue, C. Doetkott, M. Marchello, and J. Sherwood, "Independent component analysis- processed electronic nose data for predicting salmonella typhimurium populations in contaminated beef," Food Control, vol. 19, 2008, pp. 236–246.

[55] M. Kermit and O. Tomic, "Independent component analysis applied on gas sensor array measurement data," IEEE Sensors Journal, vol. 3, 2003, pp. 500–511.

iaria journals

www.iariajournals.org

International Journal On Advances in Intelligent Systems
issn: 1942-2679

International Journal On Advances in Internet Technology
issn: 1942-2652

International Journal On Advances in Life Sciences
issn: 1942-2660

International Journal On Advances in Networks and Services
issn: 1942-2644

International Journal On Advances in Security
issn: 1942-2636

International Journal On Advances in Software
issn: 1942-2628

International Journal On Advances in Systems and Measurements
issn: 1942-261x

International Journal On Advances in Telecommunications
issn: 1942-2601